Industrial Robotics - New Paradigms

Edited by Antoni Grau and Zhuping Wang

Published in London, United Kingdom

IntechOpen

Supporting open minds since 2005

Industrial Robotics - New Paradigms
http://dx.doi.org/10.5772/intechopen.83174
Edited by Antoni Grau and Zhuping Wang

Contributors
Shouren Huang, Yuji Yamakawa, Masatoshi Ishikawa, Edgar Mario Rico Mesa, Laura Cecilia Tobon Ospina, Juan David Arismendy Pulgarin, John Sneyder Tamayo Zapata, Paula Andrea Palacios Correa, Eiji Nakano, Hidetoshi Ikeda, Natsuko Muranaka, Keisuke Sato, Bin Fang, Fuchun Sun, Chao Yang, Marcello Chiaberge, Luca Navilli, Lorenzo Galtarossa, Evanthia Zervoudi, Edgar Omar Lopez-Caudana, German Baltazar, Pedro Ponce Cruz, Hong Seong Park, Sang Hoon Ji, Donguk Yu, Hoseok Jung, Antoni Grau, Huaping Liu, Edmundo Guerra, Jordi Palacin, Zhuping Wang

First published in London, United Kingdom, 2020 by IntechOpen
IntechOpen is the global imprint of INTECHOPEN LIMITED, registered in England and Wales, registration number: 11086078, 7th floor, 10 Lower Thames Street, London, EC3R 6AF, United Kingdom
Printed in Croatia

British Library Cataloguing-in-Publication Data
A catalogue record for this book is available from the British Library

Additional hard and PDF copies can be obtained from orders@intechopen.com

Industrial Robotics - New Paradigms
Edited by Antoni Grau and Zhuping Wang
p. cm.
Print ISBN 978-1-83880-733-7
Online ISBN 978-1-83880-734-4
eBook (PDF) ISBN 978-1-83880-735-1

Meet the editors

Antoni Grau received his M.S. and Ph.D. degrees in computer science from the Technical University of Catalonia (UPC), Barcelona, in 1990 and 1997, respectively. He is currently a Professor with the Department of Automatic Control, UPC, giving lectures on computer vision, digital signal processing, and robotics at the School of Informatics of Barcelona. His research interests include computer vision, pattern recognition, autonomous mobile robots, factory automation, and education on sustainable development. He has chaired several international conferences. He serves as an Associate Editor of the IEEE Transactions on Industrial Informatics.

Zhuping Wang received her B.Eng. and M.Eng. degrees from the Department of Automatic Control, Northwestern Polytechnic University, China, in 1994 and 1997, respectively; and her Ph.D. degree from National University of Singapore in 2003. Currently, she is a Professor with the School of Electronics and Information Engineering, Tongji University, Shanghai, China. Her research interests include intelligent control of robotic systems, self-driving vehicles, and multi-agent systems.

Contents

Preface

When the IntechOpen editorial team asked us to edit a book on robotics, we had no doubt about the main subject: Industrial Robotics. There are many books that covered the historic use and deployment of robots in industrial environments. Our approach is different as it is adapted to our present times. There are two main factors that distinguish the industry nowadays: the high degree of industrialization in the world; and the change to a new industrial revolution: Industry 4.0. Both factors have started an evolution in many subjects closely related to robotics: new materials, new sensors/actuators, new floor management, and new applications that present cutting edge technologies. Based on data from the International Federation Robotics IFR, worldwide robot installation grew by 6% (to 422,271 robots), representing 16.5 billion USD (without peripherals and software), and the operational stock of robots was accounted by 2,500,000 units (+15%). Even the more involved industries (customers, automotive and electrical/electronics industry) was surprised at this increase in robotic figures due to a difficult year due to the uncertainty in the global economy. However, the automotive industry is still the largest customer with a total of 30% in installed units, ahead of electrical/electronics (25%), metal and machinery (10%), plastics and chemical products (5%), and food and beverages (3%).

The new technologies involved in Industry 4.0 will reshape the way that industrial processes are done nowadays. The trend is gathering force, and company executives need to carefully monitor the coming changes and develop strategies to take advantage of the new opportunities. The trend will be driven by four disruptions: the dramatic increase in data volumes, computational power and connectivity; the emergence of analytics and business-intelligence capabilities; new forms of human-machine interaction; and improvements in transferring digital instructions to the physical world, such as advanced robotics and 3D printing. With Industry 4.0, manufacturers will be able to operate smarter factories, in which they can more easily tailor products for specific customers. While some of these technologies are not yet ready for application at scale, many are now at a point where their greater reliability and lower cost are beginning to make sense for industrial applications.

This book is intended to cover some classical topics and subjects in Industrial Robotics with the latest technologies. The eruption of Industry 4.0 brings a new kind of use of robotics in industry environments; cooperative, collaborative, and co-robot techniques are entering the floor with strength. Humans are sharing the environment with the new robot designs; security, communications and interaction, simulated and hardware-in-the-loop (HIL) are important issues. The book emphasizes the new industrial applications that emerge with the use of autonomous robots; unmanned aerial and surface vehicles are new technologies that open classical applications to be automated. Those robots need new control methodologies, and new sensoring technologies and algorithms that will also be covered in this book. Editors also give special attention to education and training in robotics, and the ethical use of the new era of industrial robotics.

The book is divided into two blocks. The first block is devoted to some new polices that have to be considered after the appearance of Industry 4.0. The article "*Fourth Industrial Revolution: Opportunities, Challenges, and Proposed Policies*" collects a set of policies together with the new opportunities and also challenges the policies that this revolution represents. In this same block, two more articles can be found. Education has a very important role in society and this new era of high technologies demand requires more pedagogical interest than ever. The article "*Training by Projects in an Industrial Robotic Application*" presents a tool to introduce and teach youngsters the robotics fundamentals that will be found in any engineering degree or simply to be introduced to this subject. In the article "*Socially Assistive Robotics: State-of-the-Art Scenarios in Mexico*", the reader will find a specific example of blending technology and culture in Mexico. This pedagogical example can be exported to any other country.

The second block of the book is devoted to applications that go beyond the classical industrial robotics. This is another strong point of this book. The new revolution, Industry 4.0, is demanding new, challenging applications with robots that are different than the classical manipulator on the floor factory. Also, the factory has to be thought as a new concept; factory can be any place where a robot can help to produce a good, covering obviously places that would not be considered real factories prior to Industry 4.0. The chapter "*Dynamic Compensation Framework to Improve the Autonomy of Industrial Robots*" covers research to enlarge the autonomy of robots in the industry which are not logically industrial manipulators. The chapter "*Cooperative Step Climbing Using Connected Wheeled Robots and Evaluation of Remote Operability*" presents a new morphology for a robot that can go through a factory with steps, this factory can be any place where a production process is happening. The chapter "*Real-Time Robot Software Platform for Industrial Application*" presents novel software that can be used for industrial robots to be executed in real-time. The next two chapters "*Visual-Tactile Fusion for Robotic Stable Grasping*" and "*Deep Learning-Based Detection of Pipes in Industrial Environments*" present applications that need external sensors such as computer vision and LiDAR; those sensors, although used in present factories, are used in new and challenging applications, the first one is for robot manipulation and grasping which is a tough task and not fully deployed in industry. The second one is a novel application to detect pipes in a large factory full of pipes (gas, oil factories for instance) that need repairing and maintaining, the detection is done with a drone able to reach complex areas of the factory. Finally, the chapter "*Visual-Inertial Indoor Navigation Systems and Algorithms for UAV Inspection Vehicles*" presents how a drone can accurately fly and locate inside environments where satellite signals do not arrive. More and more industrial applications are based nowadays on drone operations and their location has to be independent of GPS signals in the indoor environment.

The authors would like to thank the editing staff of IntechOpen for their valuable help in preparing the book, and also our colleagues at the Vision and Intelligent Systems (VIS) at Technical University of Catalonia and Tongji University for fruitfully brainstorming meetings improving the editorial line of this book.

Antoni Grau
Automatic Control Department,
School of Informatics of Barcelona,
Technical University of Catalonia, UPC,
Barcelona, Spain

Zhuping Wang
Department of Control Science and Engineering,
School of Electronics and Information Engineering,
Tongji University,
Shanghai, China

Section 1

Policies and Education

Chapter 1

Fourth Industrial Revolution: Opportunities, Challenges, and Proposed Policies

Evanthia K. Zervoudi

Abstract

In this paper, key elements about the Fourth Industrial Revolution are set under examination. Concerns, challenges, and opportunities related to the Industry 4.0 are analyzed, and specific policies to deal with the challenges and take advantage from the opportunities are proposed. Other issues that are set under consideration in this paper are the rate at which the human labor is threatened by the technological achievements, the main factors that increase workers' exposure to the risk of automation, the jobs that are more at risk due to automation, and the basic factors that make political intervention necessary in order to deal with the unpredictable consequences of the technological progress such as the threat of a nuclear disaster and a possible income and social inequality gap widening. Finally, a special reference is done for the case of Greece.

Keywords: Fourth Industrial Revolution, industry 4.0, automation, technological progress, creative disaster, robots, artificial intelligence, STEM, true creativity, social intelligence

1. Introduction

In the last decades, the technological progress was remarkable. The fast and major technological changes offer the chance to improve human life, but they also create concerns about the future. One of the biggest fears related to the new technologies is that the robots and the artificial intelligence will replace the human factor in work leading to the "technological unemployment." This is not the first time that people face the technological progress as a threat for their jobs. In the nineteenth and twentieth centuries, when another major wave of technological progress took place, similar fears had arisen, but they had not been proven right; technological achievements of these centuries finally drove to the creation of new jobs that had fully compensated the consequences of the new job-saving technology adoption ("capitalization result").

However, in view of the Fourth Industrial Revolution that has already begun in Europe and in the United States, the fear that the automation and the digitization will drive to the "End of Work" [1] wakes up again. A great discussion about the possibility of human factor replacement by machines and robots and a probable "creative disaster" have been emerged in a series of studies. Frey and Osborne [2] in their study support that 47% of jobs in the United States may be at risk of automation in the near future (see **Figure 1**). Bowles [3] in his study concludes that the proportion of sensitive-in-automation jobs in Europe varies from 45–60%,

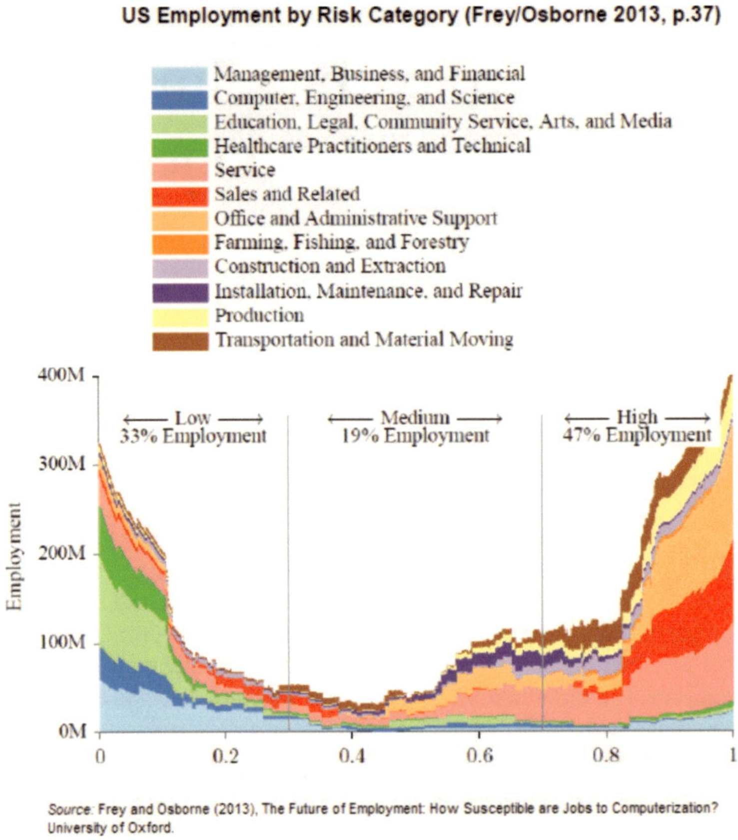

Figure 1.
Employment by risk category in US.

with Southern Europe being more exposed to a possible automation wave. The discussion about the consequences of the Industry 4.0 in World Economic Forum in Davos (2016) concluded that about 7 million jobs are at risk in the next 5 years with women being more affected.

There are various factors that could expose workers at the risk of automation. A low *work experience* is such a factor and mainly concerns young people who usually work as unskilled staff in routine positions that could be easily automated. *Low levels of education and training* is another crucial factor. Highly educated and highly specialized employees are less threatened by unemployment due to automation in contrast to low-skilled staff, whose tasks can be easily automated. The high percentages of people out of education, employment, or training (high NEET%) aggravates the situation since the difficulties of less-specialized workers to reenter into the labor market and get adapted to the new conditions will be great if they stay out of education, employment, or training for a long time. **Figure 2**[1] shows that there is a decreasing trend between educational level and the share of workers at high risk of automation; people with lower secondary education are the most exposed to the risk of automation, while highly educated employees with a Master's/PhD are the most protected against the risk of automation.

The *low degree of adaptation to automation* is maybe the most important among the risk factors of exposure to automation. Countries must acquire the mechanisms to help their citizens to be quickly and easily adapted to the new reality. In

[1] See [4].

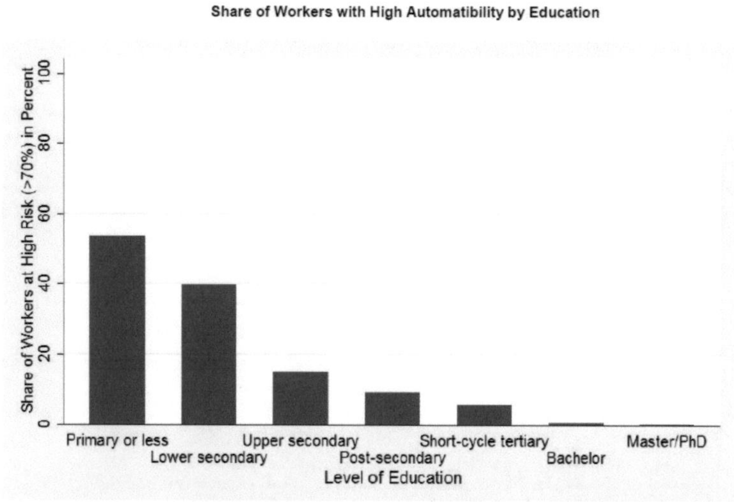

Source: Arntz, M, Gregory T. and Zierahn U. The Risk of Automation for Jobs in OECD Countries: A Comparative Analysis. (see [2])

Figure 2.
Share of workers at high automation risk by education level.

technologically advanced countries such as Japan, South Korea, and Singapore, men are increasingly working with robots in order to be highly adapted to automation reducing in this way the unemployment risk in comparison with other countries where adaptation to automation is slower.

2. Professional sectors and jobs more exposed to automation

The Fourth Industrial Revolution does not seem to threaten the human work as a whole.[2] The heterogeneity of jobs even within the same professional sector is great. Employees are differently exposed to automation depending on the position they hold and on their tasks. *Routine jobs with a high volume of tasks* related to information exchange, sales, data management, manual work, product transfer and storage, constructions, and office work are more exposed to the risk of automation. *Construction and Manufacturing* and *Wholesale and Retail Trade* are the professional sectors that are expected to be highly automated until 2030, with an estimated automation of approximately 45 and 34%, respectively (for OECD[3] countries). On the other hand, the risk of automation is lower for *jobs with high educational requirements*, the tasks of which demand *high communicative and cognitive skills*. Such tasks cannot be defined in terms of codes and algorithms (Engineering Bottlenecks); they are more related to the perception, the ability to manage complex situations, multilevel activity and flexibility, and the *true creativity*, for example, any task that cannot be provided by a machine but requires critical thinking such as the ability to develop new theories, literature, or musical compositions. There are also tasks that require *social intelligence and comprehension* such as elderly care; for these tasks there is a strong social preference to be provided by human employees and not by robots. *Health and* education are the professional sectors with the lowest estimated rates of automation (around 8–9% for OECD countries). This is also clear in **Figure 3** according to which "Transportation and storage" and "Manufacturing" are the economic sectors that are more exposed to the risk of automation (up to 50%), while sectors such as "Human health and social

[2] See [5–7].

[3] Organization for Economic Co-operation and Development (OECD).

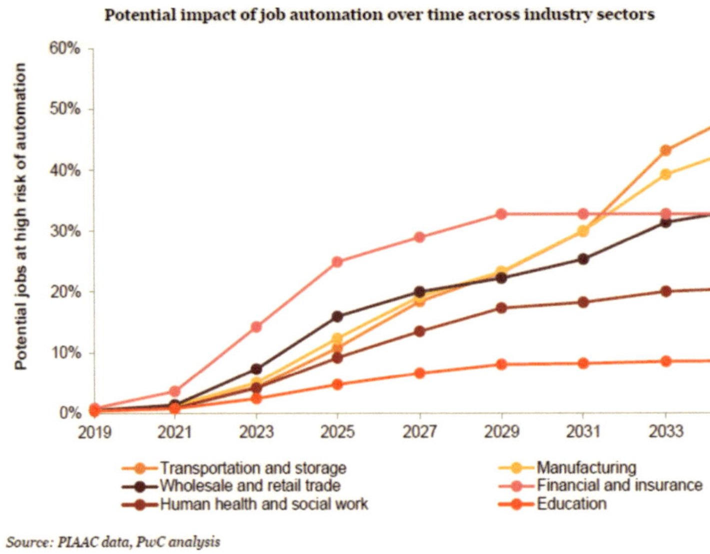

Figure 3.
Potential impact of job automation across industry sectors.

work" and "Education" are the most protected against the automation risk implying that there are tasks such as teaching and nursing that cannot be replaced by machines.

3. Challenges related to the Fourth Industrial Revolution and policies to deal with them

Major technological achievements may imply significant public policy issues. McKinsey [8] in its report underlines that the key for the successful adaption to the new technological conditions is the ability of governments to adopt the right policies. Governments that will not be able to follow the appropriate long-term policies will set their economies at risk, that is, when all the other economies will run with great speed, their inability to be adapted to the new reality will drive to the deterioration of their competitiveness, the reduction of their revenue, and the increase in their spending with the possibility of a bankruptcy to be increased. But it is not only the ability of governments to be adapted to the new conditions. There are also severe social problems that may get bigger due to the Fourth Industrial Revolution making policy intervention crucial. Political leaders must ensure that the technological progress will work for the benefit of the society and not against it. Some of the most significant challenges that may arise due to the Industry 4.0 and basic policies to deal with them are given below (see [4, 9, 10] among others). Given that the Industry 4.0 is directly related to socioeconomic growth, these policies must be in complete accordance to the Sustainable Development Goals (SGs) adopted by United Nations Member States in 2015.[4,5]

[4] https://www.un.org/sustainabledevelopment/sustainable-development-goals/.

[5] **17 Sustainable Development Goals (SDGs):** No Poverty (SDG1), Zero Hunger (SDG2), Good Health and Well-being (SDG3), Quality Education (SDG4), Gender Equality (SDG5), Clean Water and Sanitation (SDG6), Affordable and Clean Energy (SDG7), Decent Work and Economic Growth (SDG8), Industry, Innovation and Infrastructure (SDG9), Reduced Inequality (SDG10), Sustainable Cities and Communities (SDG11), Responsible Consumption and Production (SDG12), Climate Action (SDG13), Life below Water (SDG14), Life on Land (SDG15), Peace and Justice Strong Institutions (SDG16), Partnerships to achieve the Goal (SDG17).

A major area on which governments should focus is that of work. The world of work becomes increasingly complex driving to loss of millions jobs. In the EU a significant decrease in the number of low and medium skilled jobs is already observed. The use of robots significantly reduces the labor cost and the likelihood of human error, while artificial intelligence begins to substitute the human factor even in jobs that require personal contact such as sales and customer service. The World Bank [10] estimates that the increase in automation will get at risk almost 57% of jobs in OECD countries, 47% of jobs in the United States, and 77% of jobs in China. Substantial differences concerning the impact of automation on jobs are also observed among countries, for example, the proportion of workers at high risk (due to automation) in Germany and in Austria is 12%, while in the technologically advanced Korea and Estonia is 6%. However, it is a common ascertainment that in all countries, the most educated and high-skilled workforce is able to be better adapted to the new technological requirements and enjoy higher real wages, while less educated and low-skilled workers are burdened by the cost of automation, being more exposed to income loss and unemployment.

Therefore, the basic policy that governments should follow in order to reduce the risk exposure of employees to automation is the investment in education and training for people of all ages so as to be able to be better adapted to new technologies and digitization. More specifically, a government should support (i) the practical training of professionals through job-related re-skilling and up-skilling programs so as to help people to get familiar with new technologies and become more competitive in labor market, (ii) the practical education and training of children and young people in new technologies so as to enter into the labor market having the appropriate skills and the necessary knowledge, (iii) the direct connection between education and labor market, (iv) the training in STEM (Science, Technology, Engineering, and Mathematics) subject areas and the active participation of young people in such programs as young people in South Korea, Japan, Singapore, India, and China do, (v) internships and practice for young people (up to 24 years old) in order to gain work experience during their studies, and (vi) adult learning and lifelong learning programs so as to help elder people to be smoothly adapted to new technologies and digitization. Another significant goal of governments must be the job creation. The investment in education and training can be effective only if the right jobs are available. The public investment in sectors such as infrastructure and housing could benefit the long-term productivity of the economy driving to the increase of demand and the job creation.

Another issue that may arise due to the Fourth Industrial Revolution is the income inequality gap widening. Nowadays, global income inequality is at very high levels with the richest 8% of the world's population to earn half of the world's total income and the remaining 92% of people the other half. The income inequality rises globally in a fast pace. Between 1990 and 2010, the income inequality in developing countries reached at 11%. The rapid technological progress and the introduction of new technologies in all sectors, in combination with factors such as the insufficiently regulated financial integration and the growing competition in product and service markets, may widen this income inequality gap. The most educated and highly qualified staff has the ability and the skills to be better adapted to automation, and thus they will be widely benefited by the technological achievements. Moreover, people whose income, skills, and wealth are already high will be further favored by the significant increase of their assets' value because of the technological progress. On the other hand, low-skilled workers will experience unemployment and constant downward pressure on their wages and their income. The workers that will be most affected by the Fourth Industrial Revolution will be those that may now feel invulnerable to competition with robots, that is, those whose jobs require moderate skills such as customer service that could be easily replaced by artificial intelligence. Many studies and reports underline that without the appropriate policies, the Fourth Industrial Revolution may contribute to the widening of the income inequality gap with unfavorable consequences for the society. **Figure 4** below depicts this decreasing trend between income percentile and the share of workers at high risk of automation; people with lower income percentile (less than 10%) are the most exposed to the risk of automation, while well-paid employees with income percentile more than 75% are the most protected against the risk of automation. The fact that the well-paid employees are usually highly educated people highlights once more the importance of the education as a shield against the risk of automation.

Studies that are referred to the relation between the Industry 4.0 and the income inequality are that of Acemoglu [11], Barro [12], Krueger [13], Krusell et al. [14], Hornstein et al. [15], Berman et al. [16], Card and DiNardo [17], Huber and Stephens [18], and Benioff [19], which argue that technological changes affect income distribution and deepen the gap between high and low-skilled workforce concluding that the income inequality gap expansion is due to the technological crises that can disproportionately increase the demand for capital and drive to a great job loss due to automation. Birdsall [20] in his study supports that the technological progress increases the "skill bonus" and replaces low-skill workers, deepening in this way the inequality. Papageorgiou et al. [21] conclude that variables such as technological development, access to education, sectorial employment rates, and national economic growth are deterministic for inequality in low- and high-income countries. In these variables, the International Labor Organization adds the technological change, the globalization, and the reduction of social welfare as key factors for widening income inequality. An alternative point of view is that of Goldin and Katz [22] according to which income inequality is mainly explained by changes in education rather than shifts in technology. In her study, A. Guscina [23] argues that during the period of pre-globalization (pre-IT period), technological progress enforced labor reducing the income inequality, while in the post-globalization period, technological progress enforced capital increasing in this way the inequality. According to the Deloitte Global report [24], the adoption of emerging technologies as artificial intelligence in countries such as India, South Africa, and China may drive to social turmoil and increase income inequality in the future. These countries

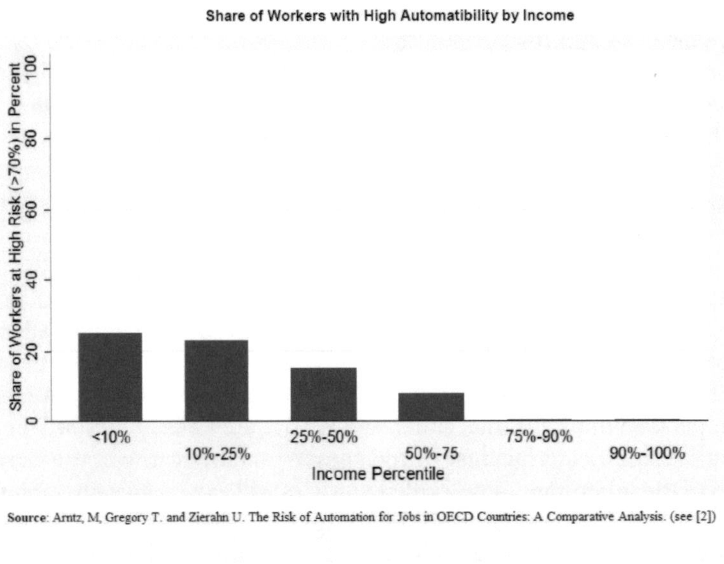

Figure 4.
Share of workers at high automation risk by income level.

had significant economic and political changes that in some cases led to high growth, but at the same time unknown "social cracks" had been introduced creating greater sensitivity to future social and economic changes. Kuzmenko and Roienko [25] in their study support that the income inequality will rapidly grow (under the influence of the Fourth Industrial Revolution) not only in the emerging economies but also in the developed countries such as France, United Kingdom, and Spain. According to the report of the Swiss bank UBS [26], the Industry 4.0 will have less impact on developed economies such as Switzerland and Singapore, but in emerging markets and especially in countries of Latin America and India, the impact of the extended use of artificial intelligence and robots will be particularly unfavorable as it will reduce their competitive advantage of low-cost labor.

Another severe social problem that is possible to get bigger due to the Fourth Industrial Revolution is poverty that is growing rapidly. Today, 767 million people live below the poverty line (with $1.90 per day). The evolution of technology and the job loss may worsen this situation driving more people to the unemployment and the poverty. The problem may become deeper if one takes into account the massive urbanization that is observed internationally. By 2030, almost 60% of the world's population will be concentrated in urban areas. The rapid population growth and the non-sustainable urbanization may cause a great rising of poverty, conflicts, high waste of resources, and severe health and food security issues. In our days, one out of nine people worldwide (795 million) is malnourished.

Thus, a general conclusion is that the Fourth Industrial Revolution may contribute to the increase of poverty and hunger and to the widening of income and social inequality with rich and high-skilled people taking advantage from the technological progress and low-paid and less qualified employees suffering a greater reduction of their income. The widening of the income gap between rich and poor countries (but also within the countries) may also lead to an increase of illegal immigration which in turn may drive to serious cultural and political conflicts. Thus, the necessity of political intervention by authorities becomes crucial in order to reduce the inequalities and the negative social consequences.

Tax transformations could help in this direction. Governments may increase their tax revenue and social security contributions by workers whose earnings (income and wealth) will increase due to the Fourth Industrial Revolution such as the high-skilled people and apply a tax relief for workers whose income will be reduced. Tax revenue may be further increased by the reinforced productivity of the economy because of the use of new technologies. These increased tax revenue may finance investments in education, training, infrastructure and in stronger social security networks for those who have great difficulty to be adapted to new technologies such as elder people. Providing equal access to high-quality education and equal opportunities to people who do not have the financial ability for training and re-training, national authorities may drastically reduce the discriminations and the socioeconomic inequality. Other sensitive social policies are the extension of the existing social security benefits and the adoption of the universal basic income (UBI) in order to protect the income of people that are hit by unemployment. Finally, governments taking advantage from the opportunities that Industry 4.0 offers may also contribute to the reduction of the hunger worldwide by promoting the sustainable agricultural production and the "smart farming," organizing food quality improvement programs for all and especially for young people using digital technology and artificial intelligence and supporting innovative ways of recycling and food waste reduction.

The risk of a gender gap expansion is another social issue that requires authority attention. In the future, industrial workforce will be mainly male, with less than 10% of European programmers being women. According to the report of the

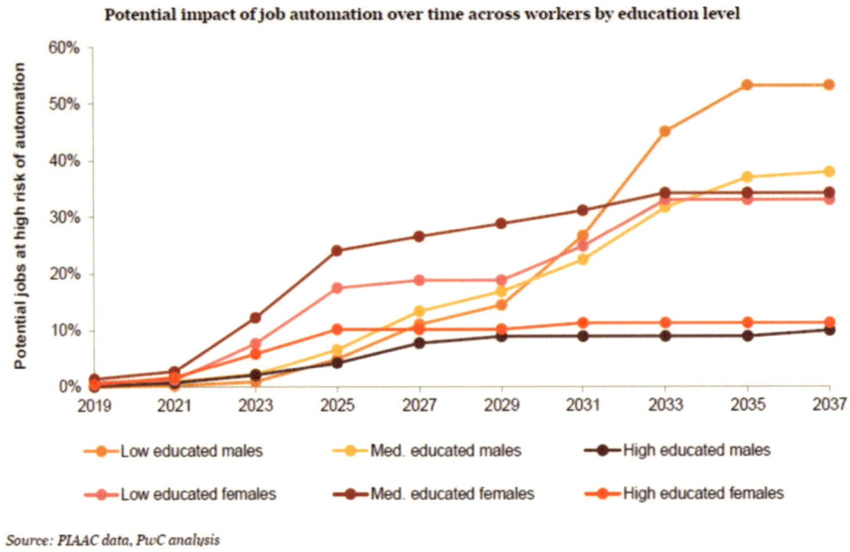

Figure 5.
Potential impact of job automation across workers by education level.

World Economic Forum, only 24% of the IT and communication sector workforce is female. McKinsey [8] in its report underlines that this fact constitutes a real business threat since companies with a higher percentage of women in managerial positions tend to perform better. Women's thinking encourages creativity and innovation and promotes the interaction between technology and society contributing to technological progress. Governments must work in the direction of addressing the gender gap by emphasizing to the female creative thinking and encouraging their active participation to the innovation processes through IT and STEM programs that will help them to become more competitive in labor market and will promote their social mobility. The protection of women's rights and the ensuring of equal opportunities for women in all countries, such as their unobstructed access to quality education, are prerequisites in order for the authorities to effectively deal with gender gap worldwide. **Figure 5**[6] captures the relation between both the educational level and the gender of employees with their exposure to the risk of automation. As it was previously highlighted, people with lower education are the most exposed to the risk of automation, while highly educated employees are the most protected ones. An interesting point in **Figure 5** is that as the automation replaces the manual work, low- and medium-educated men tend to be more exposed to automation than low- and medium-educated women, while highly educated women are constantly more exposed to automation than highly educated men but less exposed than people of low and medium education.

There are also severe legal reasons that oblige authorities to follow strict policies so as to reduce the negative consequences of Fourth Industrial Revolution for people. The transparency and the cyber security must be priorities for governments. The wide use of Internet and the increasing use of social media create the need for protection against internet bullying and personality insulting. Moreover, the great volume of personal data that is currently being collected by companies in return for providing zero-cost services obliges authorities to create strict laws and regulations that will prevent possible violations of citizens' personal data and their use in a malicious way and will protect individuals' personality. Concerning transparency,

[6] See [5, 7].

digital portals and accountability mechanisms for combating corruption may support governments' efforts and increase confidence in the governmental work. Another legal reason that requires governmental intervention is the use of new technologies for illegal activities, for example, the use of blockchain technologies for speculation purposes has been proven prone to failures and may drive to a great financial uncertainty. The use of models for secure and legal online payments and transactions and the use of new technologies for creating new, flexible, and secure service systems are crucial policies for ensuring the legality of online transactions and improving citizens' service in a safe and legal way.

The Fourth Industrial Revolution may also affect the nature of national and international security. Conflicts and wars in the new age will mainly become "hybrid" with the threat of a nuclear or chemical conflict being visible. The use of nuclear and chemical weapons in a conflict among countries requires special attention by national governments since it may cause mass destruction of populations and condemn next generations. States must proceed to strict agreements and apply the appropriate legislation in order to protect their people from the unpredictable consequences (and a probable irreversible damage) that a possible misuse of new technologies may cause on their lives and on ecosystems.

For all the above reasons, the need for cooperation among countries, at European and at international level, becomes crucial. Besides the security issues that demand the European and international collaboration in order to be addressed, such collaborations may also help countries to overcome financial and managerial difficulties that may arise at national level. The lack of interest for research and development projects by private sector (because of their great risk), the insufficient public and private funding for development projects with great social returns (because of the budget constraints), and the large funding gap in infrastructure with significant social and financial returns are important issues with which national governments may be called to deal. The coordination of national policies allows a more effective diffusion of knowledge and best practices and a more efficient use of digital innovations and country-specific business models. In this direction, governments could use new technologies to (i) promote organization and collaboration programs among businesses for information and practices' exchanging so as to increase their productivity, competitiveness, and exports, (ii) support the cooperation with European and International Institutions for funding research and development projects in all Member States, and (iii) promote the creation of forums and pan-European and international platforms so as to ensure that useful policy tools and best practices are identified, collected, exchanged, and disseminated to all countries.

Another major problem that may become more severe due to the Fourth Industrial Revolution is the climate change. Many studies have shown that the economic growth and the technological development contribute significantly to the climate change. The new species such as the drought-resistant vegetables and fruits and the new ecosystems that are created in order to deal with severe problems like hunger are up to a point helpful, but they may also affect humanity in an unpredictable and undesirable way. This fact in combination with the increasing extreme weather phenomena and the natural disasters that threaten human life (with the poorest areas to be more affected) obliges governments to take action in order to deal with climate change, sets limits in technological progress when this disturbs the environmental balance and threatens human life, and promotes the energy autonomy. In this direction, governments must use the new technologies as a tool in order to develop the appropriate policies, focusing on (i) programs and algorithms for prediction of extreme weather and climate phenomena, (ii) digital alert systems that improves the adaptability of countries to possible natural disasters, (iii) the adoption of new forms of affordable and "clean" energy such as the renewable sources of energy (wind, wave, solar) that may help countries to ensure their energy autonomy, (iv) sustainable industrialization and sustainable production infrastructure, (v) programs to promote the careful and sustainable use of terrestrial and marine ecosystems, (vi) the protection and sustainable use of forests, (vii) the protection and sustainable use of oceans and other water resources, (viii) the fight against desertification, and (ix) the protection of biodiversity.

4. The automation risk in Europe, the United States, and Asia

The estimated proportion of existing jobs at high risk of automation varies significantly by country.[7] Factors such as differences in labor market structure, education and skill levels, governmental policies on Industry 4.0, and differences in working way differentiate automation rates across countries. On the other hand, countries with similar economic structure and similar characteristics present similar potential rates of job automation (see [4, 7] among others). Four country groups that could be set under examination concerning their risk of automation are:

a. The **industrial economies**, that is, the economies where industrial production (that is easier to be automated), is still the dominant in total employment. Such economies are the **Eastern European economies** (Germany, Italy, etc.) that tend to have high shares of employment in industry sectors such as manufacturing and transport that will be easily automated until 2030s.

b. The **services-dominated economies** such as the **United States, United Kingdom**, and Netherlands, with relatively automatable jobs more concentrated in service sectors (that tend to be less automatable than industrial sectors) and low-skilled workers.

c. The **Nordic countries** such as Finland, Sweden, and Norway (in addition to New Zealand and Greece outside this region) with high employment rates, relatively less automatable jobs and high-skill workers.

d. The **Asian nations** (Japan, South Korea, Singapore, Russia, etc.) with high levels of technological advancement and education and relatively less automatable jobs but also with relatively high concentrations of employment in industrial sectors. *East Asian and Nordic economies* seem to be *less affected* by the automation (with an estimated range 20–25%), and *Eastern European economies* are *more affected* with higher potential automation rate range around to 40%, while *service-dominated countries such as the UK and US* present *intermediate levels* of potential automation. **Figures 6–8**[8] depict this potential impact of automatability across countries (individually) and across the four country groups and a range of estimates about the share of existing jobs that are at high risk of automation by the 2030s.

Eastern European countries such as Slovakia (44%) and Slovenia (42%) face relatively high potential automation rates, while Nordic countries such as Finland (22%) and Asian countries such as South Korea (22%) have relatively lower shares of existing jobs that are potentially automatable. It is important here to underline that existing jobs in some countries with low automation rates, such as *Japan and South Korea*, may face higher automation rates in the short term, given that

[7] See [4, 7].

[8] See [7].

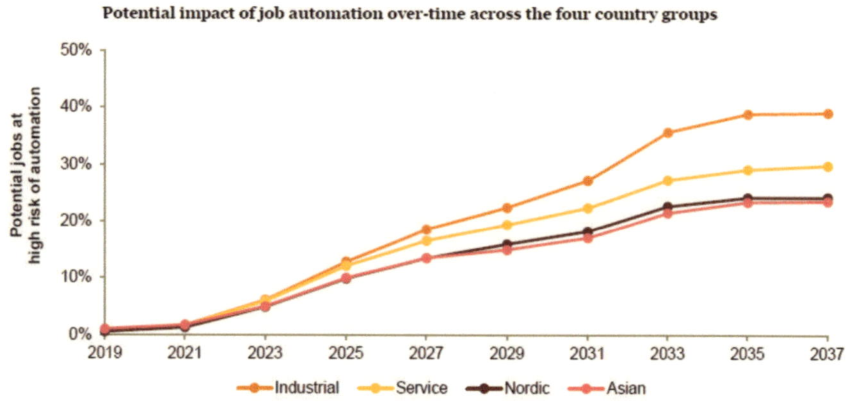

Source: PIAAC data, PwC analysis

Figure 6.
Potential impact of job automation across the four country groups.

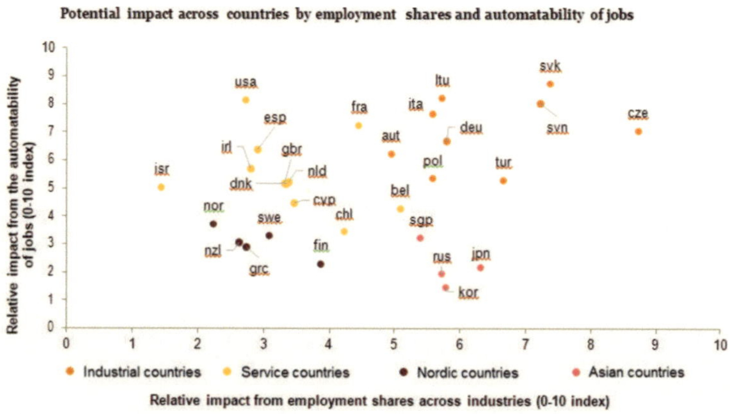

Source: PIAAC data, PwC analysis

Figure 7.
Potential impact across countries by employment shares and automatability of jobs.

algorithmic technologies are already widely used there, but in the long term (when the automation will displace manual jobs) will have lower automation rates than countries with lower average workers' skill levels and large manufacturing bases. On the other hand, countries such as Turkey may face a lower exposure in the short term but higher exposure to the later automation waves that will displace manual workers such drivers and construction workers.

Another interesting point in comparative analysis among these country groups (with an emphasis to the relation between **European and Asian countries**) is that European countries present strong negative correlations between the potential share of existing jobs at high risk of automation and the country education metrics such as government expenditure on education (as a % of GDP). This relationship is not so strong for Asian countries that present lower education spend. On the other hand, Asian countries achieve higher educational outcomes, especially in STEM subjects. Thus, the negative relationship between high education and low automatability holds for these countries as well, even with lower education spend. Furthermore, workforces in the more technologically advanced Asian countries such as Japan, South Korea, and Singapore have already adjusted to automation by *increasingly working with robots*, reducing in this way their future risk exposure

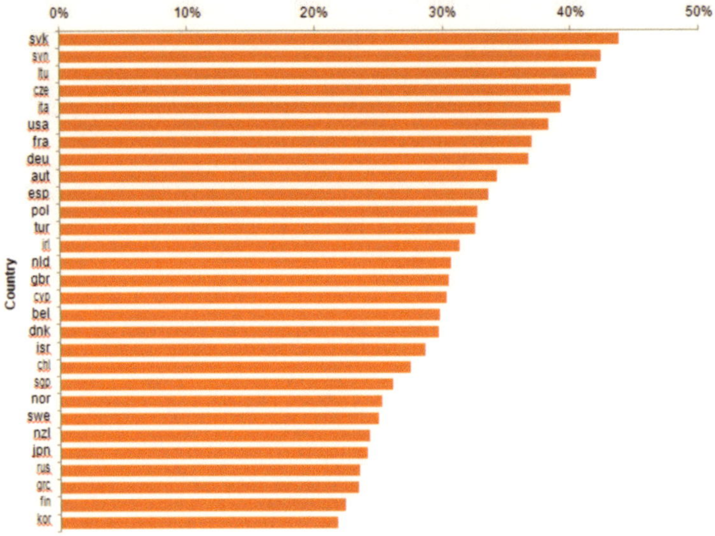

Source: PIAAC data, PwC analysis

Figure 8.
Potential rates of job automation by country.

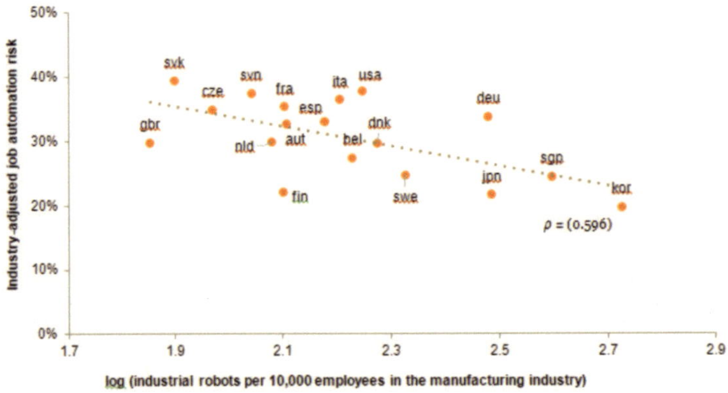

Source: International Federation of Robots, PwC analysis

Figure 9.
Relationship between density of industrial robots and industry-adjusted job automation rates.

(they may also be benefited by automation in terms of higher productivity and real wages). **Figure 9**[9] shows this negative correlation between the potential jobs at high risk of automation and the density of industrial robots per country.

Concerning the United States, a great effort has been put to integrate into the manufacturing industry the latest developments in IT, Internet, and mechanical engineering so as to reduce the risk exposure of employees to automation and get benefit by the technological achievements of the Industry 4.0. However, as Brookings Institution [27] in its report underlines the Industry 4.0, and the wider notion of advanced industries has much in common with the advanced manufacturing sector in Europe, although it includes services (e.g., software) and energy as well that led the US economy (especially services); the United States is losing ground to other countries in advanced industry competitiveness since the

[9] See [7].

labor supply, the STEM occupations, the availability of skills, and the standards in comparison with other developed countries remain poor.

4.1 The Asian giant China

The leader among the Asian countries remains **China**. China's main ambition is to become a "strong" manufacturing nation within a decade, giving priority on *digitalization, modernization, and companies' maturity in Industry 4.0 terms*, including creativity, quality benefit, and integration of industrialization, information, and green development. Two main initiatives to achieve these goals are the *Internet Plus (IP)* and the *"Made in China"* (see [4] among others). *IP* is a plan aimed at upgrading traditional industries, searching for new technologies and spreading Internet applications into the public sector, increasing both quality and effectiveness of economic and social development. *Made in China 2025 plan* is strictly focused on five major projects among which new innovation centers, green and smart manufacturing, self-sufficiency in infrastructure, and indigenous R&D projects for high-value equipment, moving industrial companies up to the value chain. The main target of the *Made in China 2025 roadmap* is to develop a domestic innovation capacity that may be been seen as *China's equivalent to Industry 4.0*: "an effort to create a manufacturing revolution underpinned by smart technologies." Moreover, a study by Fraunhofer IAO[10] about *patents registered in China* in relation to the *Industry 4.0 technologies* shows that Chinese researchers have patented important inventions in the fields of wireless sensor networks, low-cost robots, and big data, concluding that *China will be leading the pack when it comes to production data in the future*. In terms of the number of patents filed for Industry 4.0 technologies, *China has far outperformed the United States and Germany* (which is considered as a pioneer among European countries). The energy-efficient technologies intended for reliable industrial networks to robotics are basic areas in which Chinese have registered key innovations.

But the most important field of innovation in which China is considered as a pioneer among Asian countries (and worldwide) is the field of **robotics**. The number of industrial robots, using by businesses to boost their productivity, increases rapidly. According to the International Federation of Robotics or IFR (2015), the worldwide stock of robots reached in 2014 (5 years ago!) at 1.5 million units. This pace of "robotization" grows very rapidly, while the cost of new robots continues to fall and their capabilities to go up. Moreover, with the robot density in most industries to be low, the IFR anticipates that the pace of yearly robot installations will continue to grow even faster in the following years. By 2018, global sales of industrial robots were growing on average by 15% per year, and the number of units sold was around 400,000 units (see **Figures 10** and **11**) [28]. "The automation witnessed by the automotive sector and the electrical/electronics industry comes out top here with a market share of 64 percent," said IFR President Arturo Baroncelli. "A new generation of robots is a strong echo of various demands — the 'Made in China 2025' plan, US re-industrialization, Japan's rejuvenation strategy and the EU's Industrial 4.0 all symbolize the new age of equipment's transformation and a changing production mode," said Dr. Daokui Qu, CEO of SIASUN Robot & Automation. The regional breakdown reveals that 70% of the global robot sales are going to five countries: China, Japan, the United States, South Korea, and Germany. *China remains the main driver of the growth overtime and the world's biggest industrial robots market*.

[10] https://www.fraunhofer.de/en/press/research-news/2014/march/security-tools.html.

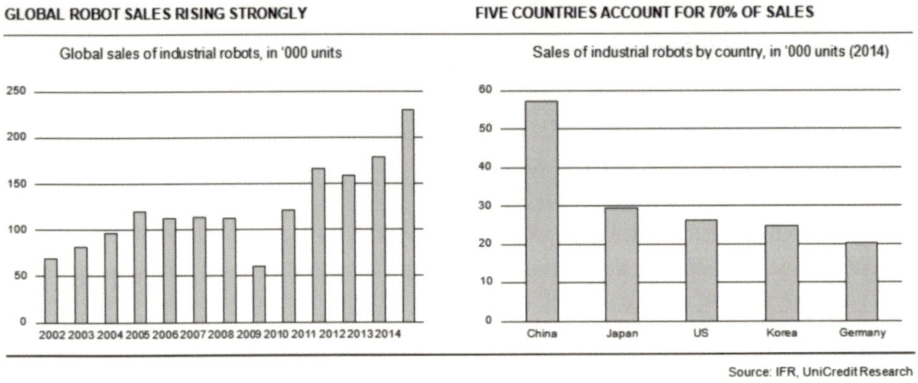

Figure 10.
Five countries account for 70% of the global robot sales that are strongly rising.

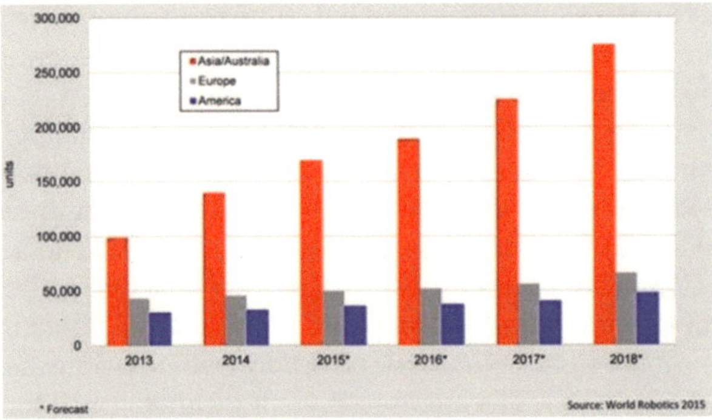

Figure 11.
Annual supply of industrial robots.

Chinese industries and country's administration have recognized the need for further automation. In 2014, sales volume reached about 57,000 units, amounted to a 1/4 of the total global sales. During 2009–2014, sales of industrial robots increased by an annual average of 59%. According to IFR "The potential remains enormous despite the recent economic downturn. Chinese production industries currently have a robotic density of 36 units per 10,000 employees. By comparison, South Korea deploys 478 industrial robots per 10,000 employees, followed by Japan (315 units) and Germany (292 units). Production industries in the United States deploy just 164 industrial robots per 10,000 employees." Statistics from the International Federation of Robotics show that China's demand for industrial robots has been growing at almost 25% per year. It is estimated that the market value in China will reach the 100 billion yuan, driving to a boom in Chinese robot manufacturers.[11] It is estimated that more than 1/3 of the global supply of industrial robots was installed in the Republic of China in 2018. China's rapid automation, says the IFR, represents a unique development in robotics history. As a result of this spectacular growth rate in robot sales, Asia, and **China** in particular, becomes the **largest and fastest growing robotics market in the world**. According to IFR, China including Japan, Korea, and other Asian countries is home to more than 60% of the robot stock in 2018, compared to 22% for Europe and 15% for the Americas.

[11] See [4].

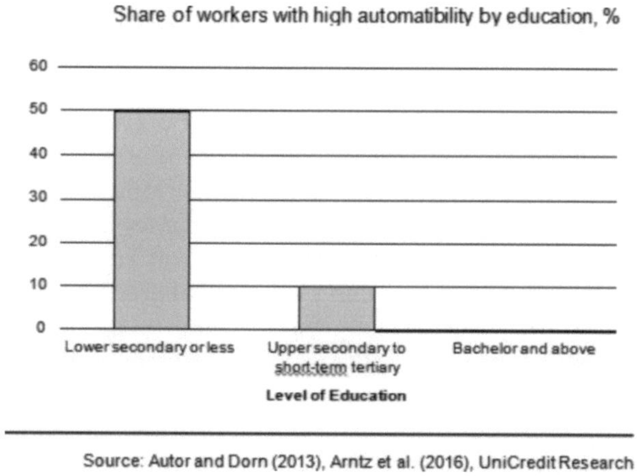

Figure 12.
Share of workers at high automatability by education.

4.1.1 The role of robotics in society

Previously in this chapter, it was analyzed the role of robots in labor market and in industry. The use of robots in industry may have both positive and negative consequences for human people jobs and lives. On the negative side, robots may be considered as a *threat for human labor* in the sense that the use of robots significantly reduces the labor costs and the likelihood of human error, and thus they may be preferred by international industries in order to reduce their costs, increase their output and their productivity, and improve their efficiency and their reliability in manufacturing by removing human errors. Moreover, job positions that were hard to be replaced by machines, such as customer services, are now easily replaced by artificial intelligence. It was also underlined that the greatest risk due to automation (including robots) is faced by low- and medium-skilled workers. The technological change overtime has been biased toward *replacing labor in routine tasks* that tended to decrease demand for low and middle-skilled occupations and increase the demand for high-educated workers rising in this way the inequality in advanced economies. *Rising inequality and slow productivity* may be the main economic challenges of the twenty-first century,[12] and the increased use of robots may affect both of these developments. There are also studies which support that robots may lift productivity, wages, and total labor demand but mostly for benefit of higher-skilled workers. In this chapter, the **great importance of education** was emphasized. As it is clear from **Figure 12** (and **Figure 2**), the risk of automation declines significantly with the level of education. Education may help people to "protect" their jobs and finally get benefit from this technological progress. Since robots are capable of taking over a great number of tasks, humans have to exploit their comparative advantages such as their cognitive skills and their capability to think out of the box in order to manage complex situations, capabilities that may be significantly strengthened by education.

On the positive side, automation may help workers to become *more efficient in their jobs* using robots as assistants/tools and entire industries and economies to become *more productive*. The productivity impact of robots is comparable to the contribution of steam engines in humanity (see [29] among others). Besides the improvement of efficiency and productivity, the use of robots in a workplace may also involve safety improvements for both employers and employees. Human workers are keeping away from dangers and risks that manual works often contain

[12] https://blogs.worldbank.org/jobs/economic-and-social-consequences-robotization.

(high risk of industrial accident) and prevent employers from potentially facing expensive medical bills and lawsuits that are always more expensive than the repair bill for a robot. Moreover, in countries where men are increasingly working with robots, their adaption to automation is easier and higher (reducing in this way their unemployment risk) in comparison with other countries where adaptation to automation is slower. Besides their impact on purely industrial activities, robots may also offer important opportunities for AI in public services such as health and social care. Smart digital assistants and intelligent robots are already valuable tools in doctors' hands in order to perform complex surgical procedures saving human lives. Robotics and AI may help to transform the whole medical ecosystem, including early detection, diagnosis, decision-making, treatment, and life care (see [7] among others). In general, there are many sectors and works where robots could be useful tools in order to facilitate people's lives and help science and humanity to go one step further. The question is whether humans are prepared for the fundamental transformation brought by artificial intelligence and automation (including robots) and whether this fundamental transformation makes social and economic sense.

In the past, radical innovations have transformed the way in which humans live together; for example, cities acquire a less nomadic character with a higher population density. More recently, the invention of technologies such as the telephone and the internet revolutionized how people store and communicate information. However, these innovations did not change the fundamental aspects of human behavior such as love, friendship, cooperation, that remain remarkably consistent throughout the world. On the other hand, the artificial intelligence and the robots' invention in our everyday life may become more disruptive. Nowadays, robots start to look and act like humans, live in our houses as personal assistants, become part of our lives, and have direct interactions with people and between each other.

The "machine behavior" is a field that does not see robots only as human-made objects but as a new class of social actors. The aspects of AI machines that should concern us are those that affect the core aspects of human social life. In 1940s, when the interaction between humans and artificial intelligence starts to seem not a distant prospect, Isaac Asimov posited his famous Three Laws of Robotics, with a main goal to keep robots from hurting people. Such a rule was "a robot may not injure a human being or, through inaction, allow a human being to come to harm." In 1985, Isaac Asimov added another law of robotics to his list: "A robot should never do anything that could harm humanity. But he struggled with how to assess such harm." "A human being is a concrete object," he wrote later. "Injury to a person can be estimated and judged. Humanity is an abstraction".

Dr. Christakis in his lab at Yale conducted some experiments in order to explore the effects of the interaction between people and robots [30]. The results were ambiguous. In some experiments, the interaction of robots with humans made people more productive and improve the way humans relate to one another, but in other experiments, the presence of robots in a social environment made people to behave less productively and less ethically. More specifically, in an experiment designed to explore how AI might affect the "tragedy of the commons," that is, "the notion that individuals' self-centered actions may collectively damage their common interests," robots converted a group of generous people into selfish persons that care only for themselves. Cooperation, trust, and generosity are key features for human social life. The fact that AI may significantly reduce people's ability to work together is extremely concerning.

There are various social effects of the use of AI in our everyday life. Many parents have noted that their children develop close relationships with AI robots and that multiple times they behave rudely to those digital assistants, that is, they give them orders in a rude way. These facts made parents to worry that this rude behavior will not be limited only to robots, but it may be expanded to the way that their kids will treat people and/or that their kids will have socialization problems in the sense that they will prefer to have relationships with AI machines instead of people. Additionally, Judith Shulevitz pointed out that as digital assistants become part of our lives, people start to treat machines as confidants or even as friends and therapists. People start to feel more comfortable to talk to devices whose responses make them feel better than to people that may hurt them. So, which is the future of human relationships? As AI become part of our lives, it seems possible for human emotions to become "something" ridiculous and the deep human relationships to be transformed into "something" superficial and narcissistic. Kathleen Richardson, anthropologist at De Montfort University in the United Kingdom and director of the Campaign "Against Sex Robots," pointed out that even love and sex will be dehumanized; the users of sex robots may pass from treating robots as instruments for sexual gratification to treat other people in the same way. Of course, there is also the opposite opinion such that of David Levy who defends in his book "Love and Sex with Robots" the positive implications of "romantically attractive and sexually desirable robots." He suggests that some people will come to prefer robot mates to human ones in sex, and this must be seen as ethical and expected since robots will not be susceptible to sexually transmitted diseases or unwanted pregnancies, while someone may easier fulfill his sexual fantasies with a robot.

Since robots are actively involved in human workplace, it would be interesting to set under consideration, besides the economic effects, the effects that such a coexistence (human workers and robots) may have on workers' psychology [31]. Of course, the overall employee psychology is affected by the robots' presence in their workplace both positively and negatively, basically depending on how the employer chooses to incorporate robots into the business. If the majority of the job positions in a workplace become automated, employees will feel insecure, unmotivated, unappreciated, and quite unhappy for the robots' presence in their workplace. On the other hand, if the robots are incorporated into the business as assistants to the current workforce, workers will feel secure and satisfied by the robots' presence in their workplace since employees will have a precious assistant to accomplish dangerous and uninteresting tasks while they will have the chance to work on more interesting and mentally stimulating tasks becoming more productive, shifting into more skilled positions and increasing their earning potential in the future.

The general conclusion is that robots and machines are already part of our everyday life, and this is a new reality that must be accepted by everyone. People must try to be adapted to this new reality in order to have a smooth transition from the old to the new world. The key is the way that people face this new reality. As it was underlined in this chapter, there are tasks such as teaching and nursing, for which there is a strong social preference to be provided by human employees and not by robots. However, robots are already used as personal assistants for elderly care with a very positive impact, for personal and domestic use and for many more categories that seem to be on the way. Based on the results of his experiments, Dr. Christakis underlined that "in what I call "hybrid systems"—where people and robots interact socially—the right kind of AI can improve the way humans relate to one another." Based on the findings of this chapter, a key word for a harmonic coexistence of robots and human people is "the right kind of AI" and the way that people treat those AI robots and machines. AI must not replace humans but they may help people to become better. AI must not be treated by humans as family members or as friends but as digital assistants that make their lives easier. In this way, people will get benefited by these technological achievements, the human feelings and the human relationships will be protected, and the genetically inherited capacities for love, friendship, cooperation, and teaching that helped people to live together peacefully and effectively across the time will not be set in danger by the AI robots and machines present in their lives.

5. Opportunities related to the Fourth Industrial Revolution

Besides the problems that may arise or get bigger during the Fourth Industrial Revolution, there are also significant economic and social opportunities that may contribute to a sustainable socioeconomic growth (see [32, 33] among others). Concerning entrepreneurship, new technologies must not be treated as a threat for human work but as a valuable tool/assistant for employees to increase their productivity and facilitate their decision-making and for entrepreneurs to boost their business competitiveness and productivity. Governments could also support entrepreneurship, focusing on the following:

i. Providing know-hows to start-ups and small- and medium-sized enterprises (SMEs) about next-generation technologies and digitalization in order to increase their revenue and reduce their production costs.

ii. Supporting co-operations among enterprises, businesses and research institutes, enterprises and people who have great market experience as business angels, businesses, and public and regional authorities.

iii. Promoting funding measures for start-ups and small- and medium-sized enterprises (SMEs) in order to help them participating in technological development processes, for example, facilitation of their access to public funding and guarantees (and to private borrowing), support of co-financing by industry and market players, and use of innovative and close-to-market financing instruments such as business loans and tax incentives.

iv. Facilitating the access to multilevel platforms that offer digital transformation programs for businesses in order to reduce information asymmetry and help businesses to remain updated and sustainable.

v. Reducing bureaucracy and barriers for business to be expanded in new markets and diversify their activities.

These policies may benefit both businesses and governments; entrepreneurs will be smoothly adapted to the new technological conditions and the digitalization having the appropriate support, and governments will increase their tax revenue due to the higher labor income and the increased business gains (due to the use of new technologies that improves businesses' effectiveness). This additional tax revenue may finance higher public spending on health and education and support additional jobs in these areas.

The new IT systems may also give to entrepreneurs the chance to participate in new supply chains for small- and medium-sized enterprises and have access to new product and service markets that under other conditions would be difficult and costly. The development of new markets with greater quantity and variety of products and services, and eventual lower prices, in combination with the improvement of the existing jobs' efficiency and the improvement of customer service, will benefit consumers driving to a demand increase and consequently to a labor demand increase. New technologies may further increase the labor demand by creating new, stable, and well-paid jobs in innovative technological sectors that will reduce the potential job loss due to automation and will substantially contribute to the fight against poverty worldwide. A characteristic example is the information and communication technology (ICT) sector that has been a key driver of economic growth in OECD countries and led to a 22% increase in jobs in 2013. Briefly, new technologies may contribute to the reduction of unemployment, to the fight against poverty and to the improvement of the quality and the prices of products and services offered to people improving in this way the quality of their lives.

In the direction of human life quality improvement, significant steps have also been done in the health sector. The broad technological innovation in the field of medicine, involving nanotechnology and genetic engineering, allow the treatment of devastating diseases and illnesses increasing the life expectancy. Moreover, smart digital assistants and intelligent robots are able to perform complex surgical procedures that under different circumstances would be impossible to be done. Except the physical health, the opportunity for more flexible forms of work due to the technological progress improves the mental health of people as well; workers have the possibility to distribute their time according to their needs, to create family and acquire a healthy social life having a better work-life balance.

Digital technology also facilitates the access of all people (in developing and developed countries) to education giving them the chance to improve their knowledge and their skills by attending educational and training programs by distance. In this way, the barriers in access to quality education for all are reduced, and the fight against inequalities and discrimination among countries and social classes becomes more effective. Moreover, the improvement of their skills enforces the self-confidence and the competiveness of individuals in labor market, helps them to be smoothly and quickly adapted to the new conditions, gives them the incentives to live and work in their country (and not to immigrate), and helps them to efficiently deal with their economic problems by becoming more productive in their work. In this way, labor income increases contributing to the reduction of poverty and hunger.

The *fight against poverty and hunger* is also supported by the technological progress in the field of sustainable agricultural production and the "smart farming," using new effective "smart" cultivation systems that may help people not only to have food for a specific period but also to learn how to easily and effectively cultivate the land in order to ensure their food forever. In this direction, the varieties of drought-resistant vegetables and fruits that may ensure food to people who live in countries that are strongly affected by drought like many countries in Africa also contribute. The technological innovations in recycling for industry and households such as the innovative composting methods may also help in the direction of food waste reduction contributing further to the fight against hunger.

6. Case study: Greece

An interesting case study is that of Greece. It is about a country that does not belong to heavy industrial economies, such that of Germany, Slovakia, and Italy which have relatively inelastic labor markets and large tertiary service sectors that may be strongly affected by the Fourth Industrial Revolution. Jobs in Greece are more related to tasks that require the involvement of human factor such as teaching and elderly care and less to routine tasks.

In general, the automation process involves three overlapping waves: (i) an *Algorithm wave* that mainly focuses on automating simple computational tasks such as structured data analysis and mathematical calculations, and it is expected to reach its full maturity by 2020, (ii) an *Augmentation wave* that focuses on the automation of repeating tasks such as communication and information exchange and statistical analysis of unstructured data that is in progress, and (iii) an *Autonomy wave* that focuses on the physical and manual work automation, such as manufacturing and transporting, that is likely to reach its full maturity by 2030.

Based on international studies' results (see [5–7], among others), less than 5% of the jobs in Greece is exposed to the automation risk due to the Algorithm wave, 10% is exposed due to the Augmentation wave and 10% of the jobs is exposed due to of the Autonomy wave, completing a percentage of about 25% of jobs in Greece that is exposed to the automation risk. This is the fourth lowest percentage of exposure to the automation risk among other OECD economies, along with some technologically advanced countries in East Asia and Scandinavia (20–25%).

Moving to an in-depth analysis of the data about the long-term impact of automation in Greece and making a separation by gender, age, educational level, and industry, one may firstly observe that the proportion of men exposed to the risk of automation (27%) is higher than that of women (18%). This is basically related to the nature of the tasks that men undertake, for example, manual work and tasks that require muscle strength and can be easily automated. Additionally, PISA scores show that women in Europe achieve better educational results than men, which may further explain the lower rate of exposure to automation risk for women. It is noteworthy that the percentage of women exposed to the risk of automation in Greece is among the lowest in Europe.

Focusing on the age groups, the highest rate of exposure (25%) is observed for the *middle-aged group (40–50 years)* and the lowest (19%) for the age group of young people with elderly people to follow with 20%. In the most European countries, the highest rate of exposure is observed for the age group of elderly people. This is mainly explained by the difficulty of elder people to be adapted to the new conditions and by the low participation rates of elderly people in labor market and in re-training programs that could help them to be adapted to the new reality. The high rates of "Not in Education, Employment, or Training" middle-aged people in Greece and their very low participation in re-skilling and up-skilling programs in order to get familiar with new technologies and become more competitive in labor market may offer an explanation for the high rate of exposure to the risk of automation for the middle-aged group in Greece.

Concerning educational level, the lowest rate of exposure to automation is observed for highly educated people (10%), the highest (30%) for people who have medium educational level, while people with low educational level present a rate of exposure of about 24%. It is about an expected result since highly qualified and educated people are at lower automation risk than medium- or low-qualified workers because of the nature of the tasks they undertake that is more complex and demanding and thus more difficult to be automated. The fact that the highest rate of exposure to automation is observed for people with medium educational level is in accordance to the results of several studies, such that of UBS [26], according to which the greatest impact of the Fourth Industrial Revolution will be experienced by the medium-skilled employees in jobs such as the customer service that although require communication skills and personal contact with the clients can be easily replaced by artificial intelligence.

The industry that appears to be most exposed to automation in Greece is the manufacturing sector with 35% rate of exposure (the 4th lowest percentage among other OECD countries). The second most exposed industry is the construction sector with 25% rate of exposure (second lowest), followed by retail trade with 23% (third lowest), social protection and health industry with 20%, and the education sector with the lowest rate of exposure of 3%. Humanitarian activities such as social protection and care services, education, and teaching require high social and cognitive skills, personal contact, and communication skills and exhibit low exposure rates to automation in comparison with the manufacturing and the construction sectors. This is in accordance to the previous findings for the Fourth Industrial Revolution concerning the sectors that are more exposed to automation. In general, the rates of exposure to automation for all professional sectors in Greece are among the lowest in Europe; especially the risk of automation of the educational sector in Greece is lower than the average of all countries worldwide, emphasizing the anthropocentric nature of the Greek educational system that makes quite difficult the total replacement of human factor by machines and robots in the long run.

7. Conclusion

Major waves of technological progress such that of Fourth Industrial Revolution always create concerns about the future of human labor and the possibility of substitution of the human factor by machines and robots. The main findings of this paper show that the Industry 4.0 does not seem to threaten the human labor under the conditions that employees are able to be quickly adapted to the new reality and governments follow the appropriate policies to protect people from the unpredictable and undesirable consequences of technological progress. The jobs that are most exposed to automation are the routine jobs with a high volume of tasks that do not require high communicative and cognitive skills such as office work, constructions and manufacturing, and wholesale and retail trade. On the other hand, jobs such as teaching, nursing, and elderly care that are multitask and require flexibility, true creativity, and social intelligence are difficult to be automated. Therefore, the complete substitution of human workforce by robots in labor market is extremely unlikely to happen.

Deloitte's report [24] characterizes the Fourth Industrial Revolution as "a mixture of hope and doubt." On the one hand, new technologies create opportunities for sustainable economic growth and reduction of unemployment; create new job positions in innovative sectors; contribute to the strengthening of competitiveness and productivity of workers and businesses, to the increase of labor income and business gains, to the improvement of human life quality, and to the physical and mental health improvement increasing life expectancy; allow for high levels of innovation and knowledge; facilitate the access to quality education for all; and contribute to the early diagnosis of extreme weather events, to the sustainable urbanization, and to the fight against inequalities, poverty, and hunger. On the other hand, the loss of millions jobs due to automation, the invasion of artificial intelligence even in jobs where the human factor is crucial, the potential income and socioeconomic inequality gap widening with the poor and developing economies to be more affected, the gender gap expansion, the increase of poverty and hunger because of the potential job loss, the violation of personal data, the use of new technologies for illegal activities, the national and international security issues such as the threat of a nuclear or a chemical conflict, and the climate change with the increasing extreme weather phenomena are some of the most important challenges related with the Industry 4.0.

Indicative key policies that governments could follow to deal with these challenges and take advantage from opportunities arising from the Fourth Industrial Revolution are the following: (1) give priority to the *education and* the *training* for people of all ages (with an emphasis to STEM issues) in order to obtain the cognitive and social skills required by the labor market and protect job positions from automation; (2) create *new well-paid jobs*, so as to moderate the potential job loss (due to automation) and deal with income and socioeconomic inequality;

(3) *strengthen social security networks*, especially for those who have difficulty to be adapted to new technologies; (4) apply *tax transformations* in order to increase tax revenue from workers whose earnings will increase due to the Industry 4.0 and apply a tax relief for workers whose income will be reduced; (5) *support entrepreneurship*, by giving small and start-up businesses the chance to improve their efficiency and increase their revenue using new technologies; (6) promote *women's participation in STEM programs* and activities in order to reduce the gender gap; (7) *support countries' cooperation*, for a better diffusion of knowledge and best practices among national governments; (8) give an emphasis to *transparency* through digital portals and accountability mechanisms; (9) impose strict rules to prevent the use of new technologies for *illegal activities* and protect people from a possible violation of their personal data; (10) institutionalize strict laws and regulations to protect people from a possible nuclear or chemical conflict with unpredictable consequences; (11) promote *smart agricultural production* in order to deal with hunger; and (12) support *sustainable use of resources*, *protection of ecosystems,* and new forms of "clean" energy as renewable sources of energy in order to deal with *climate change* and ensure *energy autonomy*. All the policies must be fully compatible with the *Sustainable Development Goals of the United Nations* in order to effectively deal with the challenges of the Industry 4.0 and ensure a sustainable economic growth.

Finally, the case study of Greece is set under consideration in this paper. Greece does not belong to the heavy industrial economies of Europe, but it has a more people-focused labor market. Greece has the fourth lowest rate of exposure to the automation risk (about 24%) among other economies worldwide, with men being more exposed to the risk of automation than women mainly because of the nature of the tasks they undertake that is easier to be automated, for example, manual works. According to the results, the highest rate of exposure is observed for middle-aged people who have medium educational level. The high rates of "Not in Education, Employment, or Training" middle-aged people in Greece and their very low participation in re-skilling and up-skilling programs and the fact that the tasks of medium-educated employees can be easily replaced by artificial intelligence offer an explanation for this result. The industry that appears to be most exposed to automation in Greece is the manufacturing sector. Humanitarian activities such as care services, education, and teaching that require high social, cognitive, and communication skills exhibit low rates of exposure to automation; especially the rate of exposure to the automation risk of the educational sector is lower than the average of all countries worldwide, emphasizing the anthropocentric nature of the Greek educational system that makes difficult the total replacement of human factor by machines and robots in the long run.

Author details

Evanthia K. Zervoudi
Athens University of Economics and Business, Athens, Greece

*Address all correspondence to: zervoudiev@aueb.gr

References

[1] Rifkin J. The End of Work: The Decline of the Global Labor Force and the Dawn of the Post-Market Era. Putnam Publishing Group; 1995

[2] Frey CB, Osborne MA. The future of employment: How susceptible are jobs to computerization? Technological Forecasting and Social Change. 2017;**114**:254-280

[3] Bowles J. The Computerisation of European Jobs—Who Will Win and Who Will Lose from the Impact of New Technology onto Old Areas of Employment? Brussels: Bruegel Institute; 2014. Available at: https://bruegel.org/2014/07/the-computerisation-of-european-jobs/

[4] Smit J, Kreutzer S, Moeller C, Carlberg M. Directorate General for Internal Policies Policy Department A: Economic and Scientific Policy Industry 4.0. European Union [Internet]. 2016. Available at: http://www.europarl.europa.eu/RegData/etudes/STUD/2016/570007/IPOL_STU(2016)570007_EN.pdf

[5] Arntz M, Gregory T, Zierahn U. The risk of automation for jobs in OECD countries: A comparative analysis. In: OECD Social, Employment and Migration Working Papers, No. 189. Paris: OECD Publishing; 2016. DOI: 10.1787/5jlz9h56dvq7-en

[6] Organisation for Economic Co-operation and Development (OECD). Putting faces to the jobs at risk of automation. Policy Brief on the Future of Work. Paris: OECD Publishing; 2018. Available at: http://www.oecd.org/els/emp/future-of-work/Automation-policy-brief2018.pdf

[7] PwC (Hawksworth, Berriman και Cameron). Will robots really steal our jobs? An international analysis of the potential long term impact of automation [Internet]. 2018. Available at: https://www.pwc.com/hu/hu/kiadvanyok/assets/pdf/impact_of_automation_on_jobs.pdf

[8] McKinsey. Industry 4.0 after the initial hype: Where manufacturers are finding value and how they can best capture it [Internet]. 2016. Available at: https://www.mckinsey.com/~/media/mckinsey/business%20functions/mckinsey%20digital/our%20insights/getting%20the%20most%20out%20of%20industry%204%200/mckinsey_industry_40_2016.ashx

[9] European Commission. Key lessons from national industry 4.0 policy initiatives in Europe [Internet]. 2017. Available at: https://ec.europa.eu/growth/toolsdatabases/dem/monitor/sites/default/files/DTM_Policy%20initiative%20comparison%20v1.pdf

[10] World Bank. The Changing Nature of Work [Internet]. 2019. Available at: http://documents.worldbank.org/curated/en/816281518818814423/2019-WDR-Report.pdf

[11] Acemoglu D. Technical change, inequality and the labor market. Journal of Economic Literature. 2002;**40**(1):7-72

[12] Barro RJ. Inequality and growth in a panel of countries. Journal of Economic Growth. 2005;**1**:5-32

[13] Krueger AB. How computers have changed the wages structure—Evidence from microdata, 1984-1989. Quarterly Journal of Economics. 1993;**108**:33-60

[14] Krusell P, Ohanian L, Rνos-Rull JV, Violante GL. Capital-skill complementarity and Ine-quality: A macroeconomic analysis. Econometrica. 2000;**68**:1029-1054

[15] Hornstein A, Krusell P, Violante GL. The effects of technical change on

labor market inequalities. In: CEPS Working Paper 113. 2005. Available at: https://www.princeton.edu/ceps/workingpapers/113krusell.pdf

[16] Berman E, Bound J, Machin S. Implications of skill-biased technological change: International evidence. Quarterly Journal of Economics. 1998;**113**(4):1245-1279

[17] Card D, DiNardo JE. Skill biased technological change and rising wage inequality: Some problems and puzzle. In: NBER Working Paper 8769. Cambridge, Massachusetts: National Bureau of Economic Research; 2002. Available at: http://davidcard.berkeley.edu/papers/skill-tech-change.pdf

[18] Huber E, Stephens J. Income inequality and redistribution in post-industrial democracies: Demographic, economic, and political determinants. In: LIS Working Paper Series No. 602. Luxembourg Income Study (LIS) Asbl; 2013. Available at: http://hdl.handle.net/10419/95570

[19] Benioff RM. 4 Ways to Close the Inequality Gap in the Fourth Industrial Revolution [Internet]. 2017. Available at: https://www.weforum.org/agenda/2017/01/4-ways-to-close-the-inequality-gap-in-the-fourth-industrial-revolution/

[20] Birdsall N. The world is not flat: Inequality and injustice in our global economy. In: WIDER Annual Lecture 9 [Internet]. UNU World Institute for Development Economics Research (UNU-WIDER); 2007. Available at: http://citeseerx.ist.psu.edu/viewdoc/download?doi=10.1.1.178.8962&rep=rep1&type=pdf

[21] Papageorgiou C, Jaumotte F, Lall S. Rising income inequality: Technology, or trade and financial globalization? In: IMF Working Paper. International Monetary Fund; 2008. Available at: https://www.imf.org/external/pubs/ft/wp/2008/wp08185.pdf

[22] Goldin C, Katz LF. The Race between Education and Technology. Cambridge, MA: Harvard University Press; 2008

[23] Guscina A. Effects of globalization on labor's share in national income. In: IMF Working Paper. 2006. Available at: http://www.rrojasdatabank.info/guscina2006.pdf

[24] Deloitte. The Fourth Industrial Revolution is Here—Are you ready? [Internet]. 2018. Available at: https://www2.deloitte.com/content/dam/Deloitte/tr/Documents/manufacturing/Industry4-0_Are-you-ready_Report.pdf

[25] Kuzmenko O, Roienko V. Nowcasting income inequality in the context of the Fourth Industrial Revolution. Socio Economic Challenges. 2017;**1**(1)

[26] Union Bank of Switzerland (UBS), Baweja B, Donovan P, Haefele M, Siddiqi L, Smiles S. Extreme automation and connectivity: The global, regional, and investment implications of the Fourth Industrial Revolution. UBS White Paper for the World Economic Forum. Annual Meeting 2016 [Internet]. 2016. Available at: https://www.ip-watch.org/weblog/wp-content/uploads/2017/09/ubs-vierte-industrielle-revolution-2016-01-21.pdf

[27] Muro M, Rothwell J, Andes S, Fikri K, Kulkarni SJ. What they are, where they are, and why they matter. In: America's Advanced Industries. Brookings Institute; 2015. p. 32

[28] Burns A. Uncaged. New statistics from the International Federation of Robotics illustrate converging patterns of data, efficiency and demand. Site Selection: http://siteselection.com/onlineInsider/uncaged.cfm

[29] Bandholz H. The rise of the machines: Economic and social consequences of robotization. UniCredit Global Themes Series. Economics & FI/

FX Research No. 36 [Internet]. 2016. Available at: https://blogs.worldbank.org/jobs/economic-and-social-consequences-robotization

[30] Christakis NA. How AI will rewire us. 2019. Available at: https://www.theatlantic.com/magazine/archive/2019/04/robots-human-relationships/583204/

[31] Harness J. The advantages of robotics in the workplace. 2019. Available at: https://bizfluent.com/info-8608752-advantages-robotics-workplace.html

[32] Digital Transformation Scoreboard 2017: Evidence of Positive Outcomes and Current Opportunities for EU Businesses [Internet]. 2017. Available at: http://ec.europa.eu/DocsRoom/documents/21501/attachments/1/translations/en/renditions/pdf

[33] Strategic Policy Forum on Digital Entrepreneurship. Accelerating the digital transformation of European industry and enterprises [Internet]. 2016. Available at: http://ec.europa.eu/DocsRoom/documents/15856/attachments/1/translations/en/renditions/native

Chapter 2

Training by Projects in an Industrial Robotic Application

Laura Tobon Ospina, Juan David Arismendy Pulgarin, John Sneyder Tamayo Zapata, Paula Andrea Palacios Correa and Edgar Mario Rico Mesa

Abstract

This chapter presents a case study of learning environments that generated a technical description of the reconditioning and commissioning of an industrial robotic arm, specifically from electronic control, mechanical design, and its application in kinematics and programming, as a pedagogical tool that powers education training. Topics are developed in a didactic way in the research hotbed such as the technology implemented to recondition the arm, the modifications that were made in terms of electrical and electronic capabilities, the analysis of the initial state of the existing electrical elements, the new devices to be implemented, and necessary calculations for the reconstruction and adaptation of the arm from the electro-mechanical point of view. It is actually the best way to promote research in the training of the student in the classroom, taking the initiative to access knowledge with the guidance of the teacher to understand the information related to the problem to be solved. The project method allows me to strengthen learning and especially the construction of knowledge of the dual relationship, SENA academy—business and theory—practice, as a training model.

Keywords: robotic arm, knowledge, kinematic, learning environments, research, repowering

1. Introduction

The progress of a country's industry consolidates on a technological basis made up of milling machine tools, lathes, loading robots, and robotic arms [1]. The reason for launching this statement base on the efficiency of the development of manufacturing processes since the greater the number of high-quality products produced in the shortest possible time, the lower the manufacturing cost [2]. Additionally, in [3], the development of robots for production process in the food industry is presented; in 2015 with [4, 5] the importance clarify not only of the implementation of robots in the industry but of how they integrate and interact safely with humans in different industrial processes and 2016 in [6]. The concept of hybrid equipment presented as a new form of industrial revolution through the use of robots equipped with new forms of technology and artificial intelligence to work alongside human beings.

In this same line, SENA through the GACIPE group belonging to the CTMA (Center for Advanced Manufacturing Technology), in the last 3 years has been developing different projects related to intelligent manufacturing systems, appropriating the appropriate management of the technology that usually work has been done worldwide on machine tool projects such as CNC machines and robotic arms of industrial order.

The educational approach of SENA of an institution with technical and technological programs such as SENA is education for work, and that is why the realization of projects such as these, of great importance with a high technological, scientific and research index, becomes a necessary and innovative support process that allows a student to interact with a robust element, which he may have to face in his work life, and which is not normally easily accessible to him; In most cases, these items purchased with large budgets are considered materials of unique manipulation by experts. However, through this methodology based on social constructivism pedagogy, the active participation of students from the beginning of the research process, encouraging their creativity, curiosity, and analysis capacity, allows them to achieve the structuring of possible solutions for the obsolescence of different teams they can face in their working life.

Therefore, the objective of the project was to promote the research capacity and the development of technological solutions from the repowering of the industrial robotic arm in order to allow interaction between teachers, students, and the business sector, with a view to strengthening applied research, the development of the ability to work in groups and interdisciplinarity and the promotion of a culture of learning.

2. Methodology

The project method was implemented, which consists of a training strategy that articulates a set of activities to solve one or several problems in the real context [7]. The realization and development of the project had a specific problem the repowering of an industrial robotic arm, which required the participation of some teachers, students, led by the research group with experience and knowledge in the subject of machines and tools, taking advantage of a real need that strengthens the student and future professional of SENA and count on the linkage of the company as part of the development of the research processes.

The execution of the proposal based on the principles of social constructivism in which the formulation of a solution strategy proposed by the student and the teacher plays a mediating role, the technological means are the tools for its operation and the didactic instrument is the industrial robotic arm [8].

The strategy implemented was carried out following the following: diagnostic, planning, intervention, and socialization phases.

The participants of the project were three members of the GACIPE research group (SENA Center for Advanced Manufacturing Technology). The members are the project director, and the leaders of the two research seedlings SIMAN and SIE. The members of the two seedbeds also participated. The SIMAN seedbed linked seven students of the Technology in Design of Mechanical Elements for its Manufacture with CNC Machine Tools. The SIE seedbed linked seven students of Industrial Electricity Technology. Finally, the company participated with the owner and some employees designing and building some pieces required in the project.

3. Conceptual framework

3.1 Learning environments

They are educational scenarios where favorable conditions of learning exist and develop, and that is not limited to the material conditions necessary for their implementation, but are shaped by the dynamics of the educational process from the development of actions, lived experiences, attitudes to material and socio-emotional conditions, relations with the environment and infrastructure [9].

3.2 Knowledge

It is a process of interaction between the subject and the medium, but the medium is understood as something social and cultural, which also involves the physical. The construction of knowledge is achieved from the moment the student constructs meanings in a structured environment and interacts with other people intentionally [10].

3.3 The social constructivist pedagogical model

It is a pedagogical model based on the interaction and communication of students and the collective construction of knowledge. The student is the result of the historical and social process. The language plays an essential role. The evaluation is dynamic. The teacher performs the role of mediator and supports the process based on his experience [11].

3.4 Project training

Project training classified as inductive teaching methods focused on student and active learning since the beginning of the training process. During the process proposed as a challenge, a problem stimulates student learning and gives him the responsibility before his process, requiring an effort of discussion and resolution of problems from the beginning of the project. The idea of constructivism is that in the formative process, the protagonist is the student who builds his knowledge from his own experience being more enriching [12]. This method developed in phases or stages such as: make a diagnosis, determine a problem, plan activities to solve it, execute them, and evaluate (socialize) them.

4. Project description

The equipment to be intervened is a robotic industrial welding arm of reference MR-2000 manufactured in 1995 by the MILLER company, and it has six degrees of freedom and weighs approximately 160 kg. Said arm provided its services last century and for different factors such as deterioration of its connections, the lack of maintenance, the obsolescence of its control system based on old processors, and the outdated electronics, were out of service.

4.1 Diagnostic stage

A technical diagnosis applied in the SENA facilities by research professors of the GACIPE research group, where a large part of the robot needed an update and their electromechanical systems had to be replaced; however, the diagnosis threw in front of the structure was stable and with high possibilities to work on her making some adjustments. This mechanical arm structure then becomes an excellent candidate for reconditioning and repowering devised for the project, which makes it a perfect mechanism for a pedagogical and research process.

When mentioning the possibility of integrating pedagogical processes into the project, the need for seedbeds to support the research process of the GACIPE group is emphasized, allowing the support of students from different technologies in the training areas related to the needs of the project, allowing then to implement the proposal of design and development of a new and reliable control system, by integrating a series of fully updated electrical and electronic devices, which allows us to perform a perfectly scalable assembly at industrial and plausible level as training, experimentation, and interaction tool for students from different areas of training (faculties), related to students in electronics, electricity, electromechanics, and programming.

4.1.1 Creation of the work team

In the robotic system, devices or mechanisms, such as some couplings and the entire frame, including the base, work well. However, all electronics, electromechanics, and software are obsolete. Therefore, an interdisciplinary team is needed to achieve the success of the project. Each team destined for a particular section according to their strengths. The development lines are mechanical, electromechanical control, and programming. Each team is made up of instructors and students of SENA to know first-hand what is a process of reconditioning of industrial machinery. Each team must carry out research work to apply their knowledge acquired on a real high-level project. It notes that periodic meetings of the GACIPE research group were held to make the fundamental decisions of the process.

4.1.2 Case study

The initial state of the arm is shown in **Figure 1**.

The initial state of the electrical panel and the original control are shown in **Figure 2A, B**.

Therefore, taking into account the initial situation of the robotic system, a brief diagnosis was made.

- Microprocessor-based technology of the 1980s of the last century.
- The control system is a mixture of digital and analog components. That is, it is not modular.
- The size of the power system is excessively large and heavy.
- The monitoring software is an old and discontinued version. That is compatible with the first versions of operating systems.
- The man-machine interface has destroyed half of the screen. There is no current replacement version.
- The structure and most of the mechanisms are in good condition. It should only do piece-by-piece maintenance due to its long inactivity.

Consequently, the decision is made to implement new technology (electronics, software, and mechanics) compatible with the existing structure and mechanisms for the execution of arm kinematics.

4.2 Planning stage

At this stage, the activities that must be carried out to meet the objective of the project are determined. The intention is to recover the original functioning of the robotic arm. That is why we study each technological module of the robot to have

Figure 1.
The initial state of the arm.

A B

Figure 2.
Module status: (A) control panel and (B) man-machine interface.

the detail of the problems that must be overcome. Therefore, each of the modules will be presented below.

4.2.1 Control module

The original cabinet was completely revised. The diagnosis of electronics is that the circuitry is in proper operation. However, its architecture is complex. There is a mixture of digital and analog elements. Some of those elements or components are obsolete or are outgoing. Even some of them are expensive. It is no longer being manufactured.

Therefore, the entire control cabinet was designed and constructed mainly with the objective of protecting, controlling, and indicating the status of the servo-drivers and the PC that are in charge of the control of the arm, for these different control devices were used, maneuver and protection such as selectors, breaker, electrical noise filters, thermal and overcurrent protection relays, sources, and light indicators among others.

The final control cabinet (**Figure 3**) was designed and built by apprentices and instructors who are part of the power seedbeds. Servo drivers are products offered by DMM. These products were successfully tested in previous projects.

The main components used in the control panel are: Driver DYN4-H01A2-00, Driver DYN2-TLA6S-00, and Computational Hardware 5I25 Superport FPGA based PCI.

The ignition cycle is as follows: when the main breaker is switched on, the system does not start; everything starts when the main contactor is activated, and the coil is activated from the system computer, before this, in order to start the computer. The control board has a power button on the door, this button is responsible for short-circuiting two pins of the PC board (PWR) to generate the ignition

Figure 3.
Control board.

of said card, and additionally this board has a power signal that activates the main contactor. At that time, the entire board is energized, initializing the sources, servo drivers, and other components.

4.2.2 Power module

Within the exploration and research work that was to be carried out, a technological surveillance activity was carried out looking for technologies similar to the original actuators of the arm in commercial, technical, and academic websites on the Internet with the students of electric seedbed that focused on the definition of power system and the determination of the suitable motors, in such a way that the arm meets or improves the original response level, in this particular case, six servomotors had to be selected, three of them, more robust (1 kW) to take care of the movement of the joints responsible for positioning, and three other smaller ones (0.12 kW) responsible for handling the grabber.

The servomotors were chosen to take into account the characteristics of the original engines, the main power, size, and electrical characteristics were reviewed to find a suitable replacement, as for the brand the DMM brand was selected for its reliability demonstrated in previous projects developed for the industry by the GACIPE research group. The servomotors used are Servomotor 410-DST-A6TKB and Servomotor 11A-DST-A6HKB. The servo drives used for the control of the servomotors are DYN4-H01A2-00 and DYN2-TLA6S-00. These drivers have within their basic applications, being implemented in machine tools, with RS232/Modbus RS485/CAN communication.

4.2.3 Robot programming

During the review of the robot, it could be evidenced that the software is obsolete and not functional with the current operating systems in force. As for the electronic infrastructure, both the man-machine interface and the control system are eliminated because their components have already left the commercial market.

To select the control software, a state of the art is done in academic pages and databases on the internet with the support of the students. Among the results, it was found that some of the existing programs in the medium are made by other people who, not finding a good option for this application, choose to design and build their own software as mentioned in [13–17], where there is an open architecture but with a lot of work to develop, in this search is the aforementioned LinuxCNC, a software based on the Linux operating system Ubuntu V10.04 (**Figure 6A**), which implements an open architecture for numerical control of CNC machines based on the RT-Linux kernel for real-time instruction execution. Some work done with the software shows good results as in [18]; therefore, the decision is made to select the free LinuxCNC software.

The servomotors are controlled by signals from an FPGA 5i25 card, a low-cost and general-purpose device for PCI, commonly used in this type of applications, which allows us to use pins and parallel port connectors to handle high compatibility with the most actuator motion systems becoming a reliable parallel port-type interface, works perfectly at 5 or 3.3 V, has 34 I/O bits with their respective pull up resistors, this card is installed on a motherboard GA -H110M integrated in the control cabinet, said board controls, through a hardware interface, the movement of the robot, using the free LinuxCNC software, for its robustness, additionally so that the apprentices know about it and include it in their project due to its advantages as a very versatile, reliable, and efficient software to handle CNC machine tools.

For the positioning of the robot, inductive proximity sensors (PNP LJ18A3-5-Z-HQ) have been located at the start and endpoints of each joint, information that is supplied to the LinuxCNC software.

Although this system has two sensors per joint, the arm is controlled by an open-loop control system supported by LinuxCNC software. The software initially requires the "home" sensors at initial start-up and by previously configuring the actuator step, step pulse, and motor speed. This software interprets the current position at all times according to the initial configuration. To have a correct and calibrated displacement, several trial and error tests performed by the trainees were required to move the arm as accurately as possible about the software and the instructions sent.

4.2.4 Robot structure and mechanisms

The robot presents a series of mechanical faults. The base of the structure does not have the servomotor or the couplings with its respective axis. Servo motors for grabber positioning do not have motion transmission belts. The limit switches for the different joints are not functional.

4.3 Intervention stage

For the development of the robotic arm repowering process, it was possible to develop a plan followed by an order of steps necessary for the execution of this process. Therefore, the following aspects were carried out.

4.3.1 Collection of bibliographic tools

Initially, the search was made for textual support material that includes documentation of the original robot model. Thus, there is clarity in the behavior and functions of the robotic arm.

Taking into account the information collected and the current state of the robot, it has been possible to have more clarity in the planning of the arm repowering process to achieve the objective of recovering the initial characteristics and conditions.

In addition, in this phase, there was an intense work of the apprentices of the electricity hotbed for obtaining the required information by configuring a brief state of art about publications of other works in the last 2 years, whose central theme is pedagogical robotics or educational robotics through the metasearch engine Scopus. This is how [19] refers to robotics in education as the appropriate option to increase the efficiency of the formation of research competencies in school studies; in [20], it goes further as robotics learning is considered as the pretext to involve multiple disciplines from the humanities, social sciences to mathematics and engineering that encourages creativity, while in [21], it not only states that the subject of robotics is attractive to students but also poses an evaluative methodology training for teachers taking into account the great complexity of the multidisciplinary theme of robotics is addressed by a cybernetic model of pedagogical feedback, collaborative learning, and empathy; and in [22], a model of robotics in education applied in classrooms of the local education system is presented, where we want to cover the needs of the current society and strengthen students' knowledge. Additionally, we have found a series of articles dedicated to the study of multiple works on educational robotics, pedagogical robotics, social robots, and the technologies used, reviewing different databases and metasearch engines such as web of science, Scopus, Science direct, IEEE, among others spanning periods ranging from the last 5 years to the last decade, this type of publications allows to have more condensed information on the contributions developed and show trends in the research processes [23–25], not only focus In the student's response they also discuss the teacher's role.

In this phase of the project certain findings and expressions of the seedlings were found as:

Understanding the search formula is essential to access quality information.

The Scopus metasearch engine converges the databases of the scientific world.

The selection of information is vital for the creation of a state of art of great quality.

The consolidation of state of the art depends on the number of items covered.

Review and survey articles are a fundamental pillar of art states.

These types of situations and possible conclusions determine the degree of commitment of the apprentices in their eagerness to apply the state of the art technique and access the best possible knowledge provided by the scientific world.

4.3.2 Design and manufacturing process of couplings and mechanical parts

In the initial maintenance of the robotic arm, it develops the design and modeling of each of its parts to later model each of the missing mechanical parts for the repowering process (**Figures 4** and **5**).

In this activity, the research seedbeds of the areas of electricity and manufacturing systems were the main support by disassembling the arm and identifying missing parts following the protocol found in the user manual of the robotic system.

During this phase, it was then possible to correctly develop each of the missing or next pieces to be modified by the dimensions of the new engines, thus maintaining the full functionality of the robotic arm and without compromising the movements in each of its axes.

As can be seen in **Figures 6** and **7**, the detailed design and modeling was required for each of the couplings present in the arm joints (axes), in which it is necessary to change each of the engines due to the obsolescence of its drivers, thus maintaining a control of each joint and identification in the process of advancement and mechanical development. In this aspect, the apprentices belonging to the seedlings carried out academic design and modeling exercises for each of the pieces that allowed them to strengthen their competences, an activity that was developed in parallel with the design work developed by teachers with the robot structure.

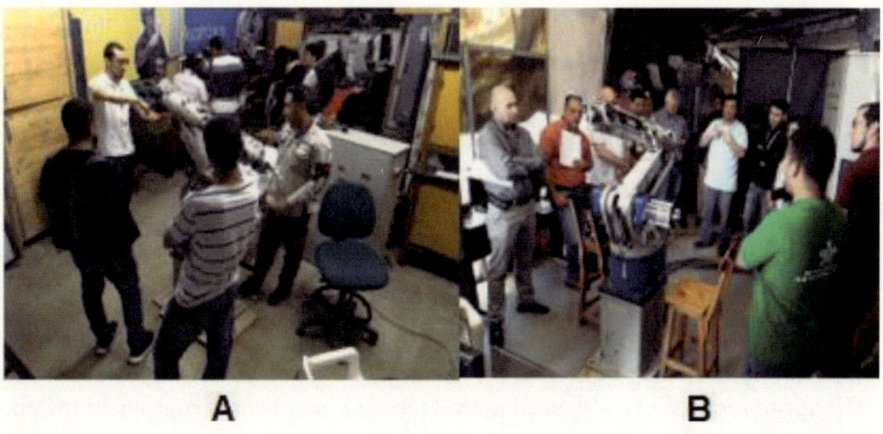

Figure 4.
Practical presentation: (A) student and (B) administrative.

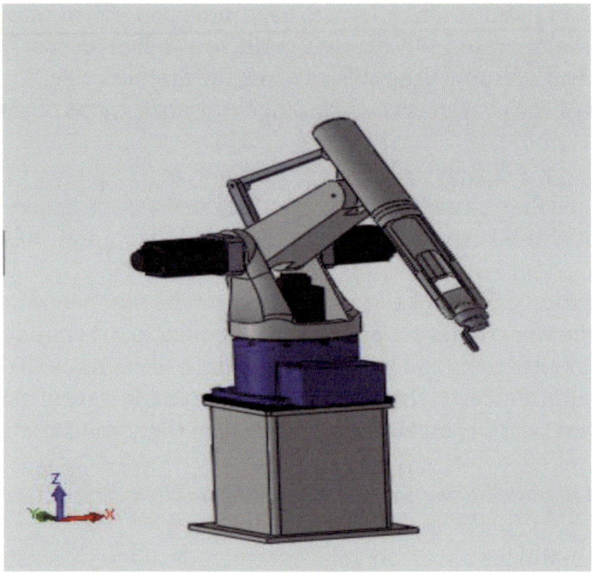

Figure 5.
Robot arm model.

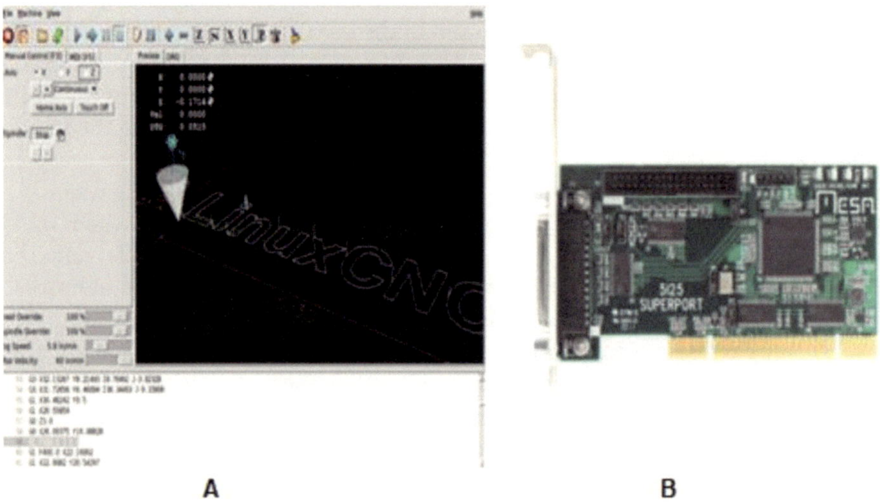

Figure 6.
Control system: (A) LinuxCNC platform and (B) FPGA 5i25 card.

Additionally, it is possible to export each of the pieces to be elaborated to a CAD program and thus start with the process of elaboration of each one of the programs for the machining routes, then the coding of its trajectories and manufacturing in machining machinery by starting of shaving (lathe, milling machine) either conventional or CNC-controlled, and the process of testing the couplings and parts in the workplace begins. In this regard, some simple pieces are developed and evaluated by the apprentices who are part of the workgroup.

In this component of the project, there are certain aspects and responses of the seedbed apprentices who showed at the meetings:

The apprentices of the advanced manufacturing seedbed strengthen knowledge in design and modeling software.

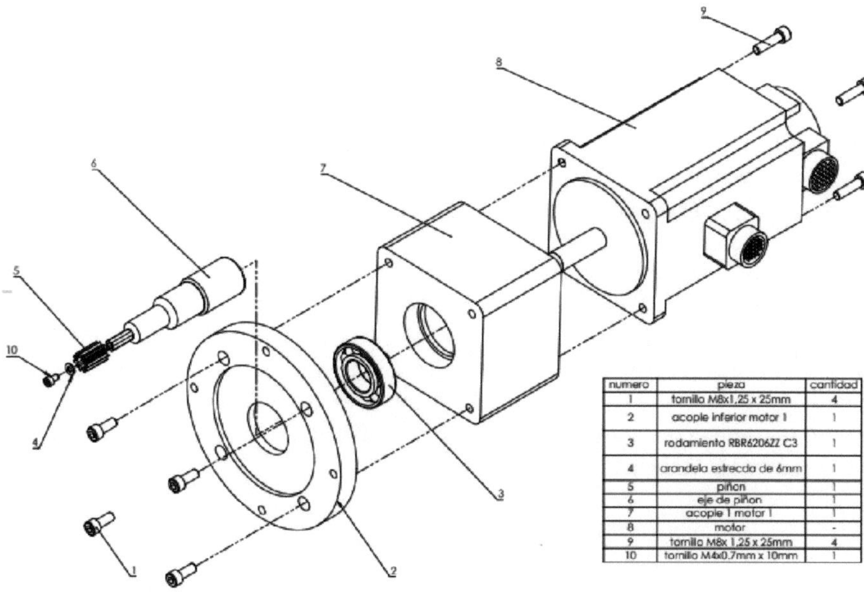

numero	pieza	cantidad
1	tornillo M8x1,25 x 25mm	4
2	acople inferior motor 1	1
3	rodamiento RBR6206ZZ C3	1
4	arandela estrecda de 6mm	1
5	piñon	1
6	eje de piñon	1
7	acople 1 motor 1	1
8	motor	-
9	tornillo M8x 1,25 x 25mm	4
10	tornillo M4x0.7mm x 10mm	1

Figure 7.
Coupling model.

When the exploded drawing of the robot was made, it was brutal to learn the different standards of the technical drawing.

The handling of machine tools has allowed us to use different metrology procedures and to know the technical norms of the manufacturing area.

The consolidation of knowledge for the use of machine tools has allowed dimensioning the procedures for machining and finishing the parts.

4.3.3 Control and power module

The technical specifications and the operating mode of both the control module and the power module are described in this section.

The board is powered by three-phase voltage, and it uses the ground and neutral signal at all times since it must power the internal lighting and the computer installed inside, both the CPU and the display (Interface). A filter is installed for each line (FN610-10-06), which provides good symmetric and asymmetric attenuation performance with high saturation resistance and excellent thermal behavior, thus having a safer system that can meet the final application need to mount on the arm (avoiding possible complications in the case of electrical noise or arc welding equipment).

This system has a fuse bank (10A-5A-2A), used to protect the board from different sides and independently. For the line that feeds the computer, an additional filter (110–220 V 6EHT1) is installed, which allows for a cleaner signal that does not affect the interaction of the arm with the computer processing. Rail terminals were also mounted (with their brakes) and a distribution terminal block for different connections such as sensors and three-phase lines, making the board more orderly and easy to review for students.

The system has protective breakers, including a three-phase, which goes directly to a contactor (three poles 240VAC-001A18NA3) that energizes the board in general. Before each servo-driver, a motor guard was installed (Gv2-me16c 9-14).

The motor guard prevents unbalance in three-phase lines. The brake of the servomotors is activated by four relays.

It also installed 3 DC sources of 48 V to 7.5 A, a dedicated source for each servodriver that controls the 0.12 kW motors, an additional 24 V DC source (for motor brakes) and a 12 V DC source (for the control).

At this stage, the electricity seedbed apprentices supported the researchers in the interpretation of plans and assembly of their components, which is how the meetings presented some findings and conclusions about the experience.

The assembly of the control cabinet is an activity for the appropriation of the technical standards of electronics and electrical.

The manipulation of the components of the module and the verification of its correct operation were the functions performed during the assembly of the cabinet.

There was care in the functional tests of the control and power module. An improper maneuver can cause the malfunction of any of the components that could lead to permanent failure.

4.3.4 Coding development robot arm

Once the electrical, electronic system, the mechanical system and the successful completion of the motion tests have been reconditioned with each independent degree of freedom, the software for the positioning of the grabber conditioned from the simultaneous movement of all the motors to this procedure known as the development of the kinematics of a six-axis industrial robotic arm [26]. In this regard, students completed training in the fundamentals of physics, geometry, and trigonometry to have the basic knowledge that allows them to understand the calculations and equations that govern the kinematics of an arm.

The position of the robotic arm is known with the mathematical expression of the inverse kinematics and direct kinematics. It should be noted that the robotic arm is a palletizer type. In databases, there is no information on kinematics for palletizers. Therefore, the kinematics of the robot was built and incorporated into the LinuxCNC.

In this regard, the students carried out state-of-the-art work related to kinematics in palletizing and serial robots to have as much information as possible to accelerate the assembly of kinematics in LinuxCNC software.

Through the LinuxCNC program, the use of the arm is allowed, either manually, independently managing each of its axes, or in a programmed way following a route or path already established by code G, thus facilitating the management process for an operator with basic knowledge in CNC coding and the area of machining.

In this process, the desired route is programmed in CAM programs, such as MASTERCAM among graduates or CNC HEEKS on the side of free programs. Initially exporting in a file (Parasolid, ages, and step), which is compatible with CAM machining programs to trace a path on which the robotic arm system should be guided.

Subsequently, the G coding is elaborated and then exported in the NGC format, with slight modifications in the program developed, so that it can be read correctly by the LinuxCNC interface and capable of being executed correctly.

In the same way, it can be programmed manually by initially acquiring the headings of the programs to follow and adding the routes in the coordinate axes of the Cartesian plane (x, y, z).

The factory application of the robotic arm was to weld parts of the automotive sector, currently, it has been repowered and conditioned to load heavy parts in the precision foundry industry.

In this phase, the students of the electricity hotbed and the manufacturing hotbed supported the design and implementation activities of the G code in the LinuxCNC to monitor the behavior of the robotic arm in its positioning in the *XYZ* plane. Additionally, the trainees started from their initiative to solve the problems, and the investigating instructors guided the students to improve the activities with the robotic arm. Then, pedagogy influences students. Therefore, the aspects that are highlighted at this stage are presented below.

The appropriation of CAM technology in free software (HEEK CNC) and licensed software (MASTERCAM) allows us to understand the scope and limitations of the coding of the instructions on the CNC machine.

The LinuxCNC application presented an appropriate performance in the robot arm.

The type of application and the type of actuator in the grabber must be established for the proper functioning of the robotic arm.

Tuning the power devices (servo drivers) of the arm prevents the crashing of your joints.

4.4 Results stage

Once the robotic arm intervention process has been carried out successfully, it has been defined to socialize the scope achieved in the execution of the project. The intention is to present an exhibition of the work teams (students and teachers) that intervened in the process of repowering by making a technical description of their contribution and finally proceeding to perform a demonstration of the robotic arm performing a welding application. The socialization executed in two moments. The first was applied during the final tests of the robotic arm, during this time six groups of students were taken, each of 20 students from different areas of knowledge to the research workshop to witness the turning of the robot (see **Figure 4A**), in addition, the administrative area of SENA was invited to verify the progress and success of the process of repowering the robotic arm (see **Figure 4B**).

Figure 8.
Presentation of the robotic arm project to the academic community.

The second moment of socialization corresponds to the completion of the tuning of the robot. The presentation in the society of the robotic system is made. The invited public is the businessmen representing Mold Glass, owner of the robot, and students and teachers from the academic community of SENA-CTMA was also invited by virtual mass media (see **Figure 8**).

During this stage, the different actors (student, administrative and businessmen) agree to qualify the results of the project as satisfactory and of great impact for the industry due to the low costs and the high degree of precision of the robot's movements.

5. Discussion

In [19], there is a collaborative work between the students and the teacher in which the researchers' terms and functions are naturally involved; in the activities proposed in the execution of the project, there are no hierarchies and each one contributes from their perception and creativity, and robotics projects are part of the curricular activities. In [20], it proposes a course by stages for the development of projects with the robotic educational analysis of state of the art, implementation, pedagogical design, learning scenarios, pilot use, validation, and social evaluation, and this procedure is performed with students of schools. The way to maintain interest in the project is through the approach of ideas proposed by students, the development of skills, and socialization of results by different means including social networks, and in [21], a conductive and constructivist activities series used to identify the benefits of the strategy called collaborative learning is executed. This procedure has been developed by university students for 5 years. As the course develops, more and more complex challenges are proposed. It is also projected towards the preparation of teachers in the management of constructivist methodologies in robotics. Therefore, a group of teachers from Public high school is trained to work with robotics projects, to apply this methodology with their high school students, but in [22], a work of systematization of subjects was done with the aim of introducing the contents on robotics from the pre-school stage to the university level. This process was carried out during 6 years that had its starting point in a summer vacation program called "Technosphere." However, despite describing and explaining the way in which these works are carried out (all immersed in the curricula) based on problems that are solved in didactic robotic systems, it is not possible to find information about the impact of these projects on life of the students because it is important to know the lived experience of the members that enriches and feeds back the constituted process, which if they are found are generic results based on surveys and evaluations that can give an indicator on the evolution of the process, but the findings obtained from interviews, logbooks and daily books are tools that have valuable information about the process. In [23–25], more general information about educational robotics works focused on doing documentary research is expressed, which indicates the importance and importance in the field of education and pedagogy since these types of treaties have within from his study trends, perspectives, and even prospects, aspects that allow validating the possibility of doing research at the frontier of knowledge.

This project describes the work done with the research seedbeds (through extracurricular activities). Applying the principles of constructivist social pedagogy using as a strategy, the project method through technological development and applied research such as the repowering an industrial robotics arm of six degrees of freedom, in which its monitoring and verification of progress, has been carried out through technical reports, the observation guide, the field diary, finding not only the technical and technological advances of the robot but also the findings, conclusions, and experiences of the work teams (teachers and students) during the development of the different stages of the project.

6. Conclusions

The development and tuning of the repowering of the industrial robotic arm allow the consolidation of a research group for the solution of manufacturing machines based on the productivity of the industry, whose technology is vital to increase its competitiveness in international markets.

This process is very important from the pedagogical point because independent of the technologies or techniques used in the project, the effectiveness of the student's learning process can be corroborated by being immersed in this type of real projects that allow to appropriate the application of the method scientific, consolidate technological knowledge and know the fundamental tool like the state of art for the development of an applied investigation.

Author details

Laura Tobon Ospina, Juan David Arismendy Pulgarin,
John Sneyder Tamayo Zapata, Paula Andrea Palacios Correa
and Edgar Mario Rico Mesa*
SENA, Medellin, Colombia

*Address all correspondence to: emrico@sena.edu.co

References

[1] Azubuike VMU. Technological innovation capability and firm's performance in new product development. Journal Communications of the IIMA. 2013;**13**(1):43-56

[2] Ben-Ari M, Mondada F. Elements of Robotics. Switzerland: Springer International Publishing; 2018. pp. 1-20

[3] Müller M et al. Robots in Food Industry Challenges and Chances. Germany: IEEE International Symposium on Robotics; 2014. pp. 1-7

[4] Tellaeche A , Maurtua I, Ibarguren A, Human robot interaction in industrial robotics. Examples from research centers to industry, Luxembourg: IEEE International Conference on Emerging Technologies and Factory Automation (ETFA); 2015. pp. 1-6

[5] Masinga P, Campbell H, Trimble JA. A framework for human collaborative robots, operations in South African automotive industry. Singapore: IEEE Advances in Intelligents Systems and Computing; 2015. pp. 1494-1497

[6] Richert A, Shehadeh M, Plumanns L, Schuster K, Jeschke S. Educating Engineers for Industry 4.0: Virtual Worlds and Human-Robot-Teams Empirical Studies Towards a New Educational Age. Abu Dhabi: IEEE Global Engineering Education Conference; 2016. pp. 142-149

[7] Kemal TA, Huseyin K. The epistemological perspectives on action research. Journal of Educational and Social Research. 2013;**3**:81-91

[8] Palacios Correa PA, Rico Mesa EM. Estudio de caso del ambiente de aprendizaje generado en un curso de introducción a la robótica. Revista Virtual Universidad Católica del Norte. 2016;**47**:173-190

[9] Villamizar LJ. The effects of Vygotsky's sociocultural theory on second language acquisition and language Input. Revista Docencia e Investigación. 2017;**7**:91-102

[10] Vera J-S, Holbrook M. Sociocultural approaches to learning and development: A Vygotskian framework. Educational Psychologist. 1996;**31**:191-206

[11] Amineh RJ, Asl HD. Review of constructivism and social constructivism. Journal of Social Sciences, Literature and Languages. 2015;**1**:9-16

[12] Prince MJ, Felder M. Inductive teaching and learning methods: Definitions, comparisons and research bases. Journal of Engineering Education. 2006;**95**(2):123-138

[13] Liu T, Wang Y, Fu H. The open architecture CNC system HITCNC based on STEP-NC. Dalian: IEEE Proceedings of the World Congress on Intelligent Control and Automation (WCICA); 2006. pp. 7983-7987

[14] Wu H, Zhang C, Li G, Wang B. Windows XP embedded based open architecture computer numerical control system. In: 2nd IEEE/ ASME International Conference on Mechatronics and Embedded Systems and Applications; 2006. pp. 1-6

[15] Huang H, Chi G, Wang Z. Development of a CNC system for multi-axis EDM based on RT-Linux. System Based on RT-Linux. In: WRI World Congress on Software Engineering; 2009. pp. 211-216

[16] Rodrigo B, Márcio F. Design of a PCB milling machine. ABCM Symposium Series in Mechatronics. 2012;**5**:1339-1348

[17] Grigoriev SN, Martinov GM. Research and development of a cross-platform CNC kernel for multi-axis machine tool. Procedia CIRP. 2014;**14**:517-522

[18] Marcos L, Iñaki D, Juan GJ. A robotic bone drilling methodology based on position measurements. In: The Fourth IEEE RAS/EMBS International Conference on Biomedical Robotics and Biomechatronics; 2012. pp. 1155-1160

[19] Skurikhina Julia A, Valeeva Roza A, Khodakova Nina P, Maystrovich Elena V. Forming research competence and engineering thinking of school students by means of educational robotics. EURASIA Journal of Mathematics, Science and Technology Education. 2018;**14**:1-16

[20] Damaševičius R, Maskeliūnas R, Blažauskas T. Faster pedagogical framework for steam education based on educational robotics. International Journal of Engineering & Technology. 2018;**7**:138-142

[21] Mario M, André KT, da Fontoura PA, Celso SK, Gustavo FD, Tiago RO, et al. Cooperative learning in robotics teacher formative assessment strategy integrated with Brazilian National High School Exam—ENEM. In: 2018 Latin American Robotic Symposium, 2018 Brazilian Symposium on Robotics (SBR) and 2018 Workshop on Robotics in Education (WRE); 2018. pp. 577-584

[22] Cheremukhina PS, Shumeykob AA. Educational robotics as a factor in the development of network interaction in the system of engineering training. Integración Educativa. 2018;**22**:536-550

[23] Souza Isabelle ML, Andrade Wilkerson L, Sampaio Lívia MR, Araujo Ana Liz Souto O. A systematic review on the use of LEGO R robotics in education. In: 2018 IEEE Frontiers in Education Conference (FIE); 2018. pp. 1-9

[24] Violeta R, Istenič SA. A review of social robots in classrooms. Emerging educational technology and teacher education. Journal Educación y Autodesarrollo. 2019;**14**:93-106

[25] Bezerra Junior JE, de Lima RW, Queiroz PGG. A study of the publications of educational robotics: A systematic literature review. IEEE Latin America Transactions. 2018;**16**:1193-1199

[26] Kucuk S , Bingul Z. Industrial Robotics: Theory, Modelling and Control Germany: Ed Sam Cubero; 2006. pp 117-148

Chapter 3

Socially Assistive Robotics: State-of-the-Art Scenarios in Mexico

Edgar Lopez-Caudana, Germán Eduardo Baltazar Reyes and Pedro Ponce Cruz

Abstract

In this chapter, we describe the experience about the use of a humanoid robotic platform, in scenarios such as education and health in Mexico. The results obtained are commented on through the perspective of cultural, technological, and social aspects in the frameworks of education (from elementary to high school) and training of health professionals. The opening towards humanoid robotic systems in elementary school children, as well as health professionals, is not far from the acceptance due not only for the technological advancement but also for different social aspects. These two considerations influenced the results obtained and experiences achieved. At the same time, this chapter shows how humanoid robotics has functioned as a tool for final projects of undergraduate students.

Keywords: social robotics, assistive education, higher education, educational innovation

1. Introduction

The implementation of biped and humanoid robots has increased during the last decade around the world. Depending on its characteristics, these robots can be implemented in different areas and domains, such as manufacture, agriculture, health, and education, among others. For the rest of this chapter, every reference will treat the last two applications mentioned, particularly rehabilitation, patient assistance, and elementary children's education.

Depending on the functionality of the robot, it can be classified into three different types [1], as seen in **Figure 1**. The first type of robot is called an assistance robot (AR). Its functionality is based only on giving physiological assistance to a patient that presents a physical disability or is recovering from a surgical operation. This type of robots regularly presents a basic structure, since they have one single task and the work environment does not vary too often.

On the other hand, as the name suggests, socially interactive robots (SIR) have the task of interacting with the end user with the sole purpose of generating a certain level of communication and entertainment to him. Generally, these platforms have a more complex design, since they need to detect and use human behavior patterns to achieve an efficient interaction.

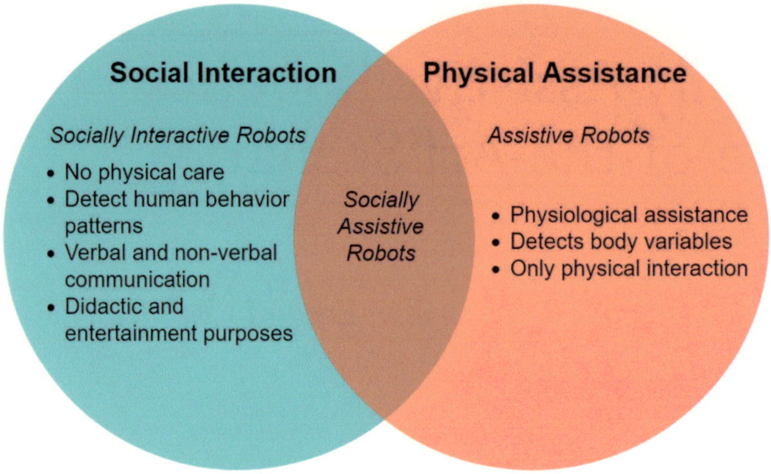

Figure 1.
Classification of robots based on their functionality.

Finally, the combination of the two prior types is called a socially assistive robot (SAR). These robots use close and empathic interaction with the user to assist in some didactic, educational, or rehabilitation process.

SAR robots are one of the most complex types of robot since it needs to resemble human behavior as much as possible to create the image of a personality and human-like behavior [2]. These two objectives allow the platform to generate empathy with the user and develop a more efficient communication with him. Also, by adequately reacting not only to the person but the environment as well, the robot is capable of performing multiple tasks.

Among all the different applications where robotics has been implemented, healthcare could be considered the area with the most impact on its use. It is usually divided into two categories: clinical and assistance. The purpose of the robots in assistance consists of the care of post-operation to help with the patient's recovery, or even to work as a substitute for a missing function in the patient's body. Another area of interest inside the healthcare is more focused on giving support to the patient on routine activities.

In the educational field, it has been recently considered that class methodologies should have an immediate application to daily life [3]. However, it is common that teachers do not know how to create innovative activities and promote student participation. One way of dealing with this problem has been with the implementation of information and communication technologies (ICTs) to improve student inclusion and link class topics with a modern approach [4]. Unfortunately, not every docent has enough knowledge on how to use these new technologies, and the new technologies turn into distractors for the students. It is then necessary to train the teaching staff into better use of the platforms to achieve better results in class.

The following sections will describe the current situation in Mexico regarding healthcare and education and how the implementation of SAR technology has been capable of improving the quality of life of the research subjects.

2. Social situations in the Mexican population

2.1 Assistive healthcare situation in Mexico

According to the National Institute of Geography and Statistics (INEGI) of Mexico, approximately 6% of the population lives with a mental or physical disability [5]. Every person that presents a physical, mental, and sensorial deficiency can be considered with a temporary or permanent disability. These disabilities limit their capabilities for performing one or more daily life routines and can be aggravated by social and economic factors.

There are multiple types of disability that affect different parts of the body. Generally, they are classified into three categories [6]:

- Sensorial and communication. Includes sight, hearing, and speaking disabilities.

- Motor. Consists of problems related to walking, manipulating objects, and coordination.

- Mental. It covers learning and socializing difficulties.

From these three categories, the motor disability is more frequent on the economically active population, while the other two types are more prone to appear in younger people [5].

Occupational therapy is defined as a client-centered health profession that promotes health and well-being through occupation, enabling people to participate in daily life activities [7]. At the same time, the World Federation of Occupational Therapists (WFOT) states that the use of assistive technology promotes independence in the patient, facilitating his participation in daily life routines [8].

In recent years, the implementation of socially assistive robots has been more frequent in Mexico, particularly during therapies with children with autistic specter, older people, and people with a motor disability. During these therapies, the behaviors of both patient and robot are observed during the performance of different occupational activities. Unfortunately, despite the promising results, there are still difficulties with a broader implementation due to economic restrictions.

Another occupational therapy study was held by the Physics and Rehabilitation National Medical Center. In this case, they studied how to detect patients with a vascular brain disease and determine if he was a viable candidate for robotic rehabilitation [9].

2.2 Educational perspective from elementary to high school level

Mexican education is divided into four different levels, each with its competences and topics to teach to their students [10]. For every level, there are specific topics regarding reading, mathematics, and science that every student needs to develop in society and further education levels. The primary purpose of the class methodologies applied is to develop the student's creativity and logical thinking, allowing them to structure their ideas and arguments [11].

Despite the efforts of the Mexican government to improve education levels, its performance on those three levels has been inefficient during the last years. The results of the PISA test held by the Organization for Economic Co-operation and Development (OECD) showed that Mexico has a deficient performance compared to other nations. From 77 evaluated nations, Mexico is located at 52nd, 59th, and 55th place in reading, mathematics, and science, respectively [12].

On national evaluations, it was also observed that 62% of Mexican students have a deficient performance in language and communication tasks, while 89.5% fail on mathematics [13].

Even though the use of ICTs has been implemented in several schools, there is still the necessity of training instructors to understand how to use these new technologies properly. Depending on the scenario, level, and topics to teach, there are different ways on how to operate robotic platforms as a supportive tool for the teacher.

3. Socially assistive robotics for elderly people

3.1 Detection of physical and psychological violence

Even though the population above 75 years old has increased during the last years, there have not been enough resources to ensure a proper healthcare system in Mexico [14]. The first significant problem is that there are not enough specialists for treatment, being only one specialized geriatric for every 10,000 elders [15]. Also, it is estimated that 10.3% of older people suffer from one or more types of abuse [16].

The work described in [17] implemented a robotic platform that gave a specialization lecture to medical students about the proper care of elder patients, required during their treatments, as well as helping to detect when they present some level of abuse. This study evaluated how volume modulation, movement correction, and body language affected the effectiveness of the robotic platform when interacting with a group of medics. **Figure 2** shows an example of an informative session.

An evaluation metric was given to the students after the session to compare the performance of the students that interacted with the robot with a different group that received the same training session from a specialized geriatric. After each session, a satisfaction survey was given to every student to evaluate as well their perceptions about the use of the platform.

The results obtained from this study showed no significant difference between the evaluation and perception results obtained from the group that interacted with the robot and the regular group. As strange as it sounds, this is an excellent result since it can be concluded that a robotic platform is equally as good as a professional when explaining specific content to a group of people. It would still be necessary to control some environmental characteristics to make sure the results are as expected. However, it opens the doors for implementing these platforms in specialized training environments.

Figure 2.
Use of a robotic platform for medical students training.

4. Socially assistive robotics in education

4.1 Use of assistive robotics for physical education classes

According to the Mexican Institute of Public Health, there has been an increment of children that suffer from obesity. From 34.8% of the population between 5 and 11 years that presented that problem, the rate of children increased to 35.6% in 2018 [18]. There has been an urgency to promote physical education (PE) in elementary schools to reduce these statistics. Nevertheless, children are not interested in those classes since the methodology is tedious and does not attract their attention.

A research group used a NAO robot as a supportive tool for the teacher to increase the enthusiasm of children during physical education class [19]. The idea was to use the platform to give the instructions to the students, as well as making little demonstrations of how they needed to do the exercises.

The experimental methodology followed the standard model of a PE class, starting with basic warm-up exercises, such as rotating arms and legs, and continuing with more intense activities, like running around an area or doing squats. Finally, the class would end after a relaxation period, where both the robot and students performed different stretching exercises, as seen in **Figure 3**.

The results of this experiment showed that the robotic platform was capable of improving the attention span of the students during class while assuring that the children performed better every one of the exercises. However, it was also shown that the intervention of the instructor was needed on certain occasions since the students did not saw the platform as an authority figure. Despite the students considering the robot as a toy, the results were positive, considering that the purpose of PE class was to make more dynamic.

4.2 Science and mathematics education in elementary school

A critical feature that elementary school needs to develop on their students is the ability to use mathematical and logical thinking in daily life situations. For this

Figure 3.
Physical education class with the support of a robotic platform.

reason, a NAO robot was used to create brief experiments regarding whole number fractions, sound propagation, and operations with the metric system [19].

The idea of the experiments was that the explanations and activities could be given by a robotic platform or by the teaching staff. Also, an observation protocol was implemented to analyze seven characteristics of student behavior:

1. Concentration (precision)

2. Concentration (recall)

3. Habituation

4. Non-habituation

5. Distraction

6. Task enthusiasm

7. Task motivation

For every session and experiment, the students were required to follow the instructions given by an instructor or the robot, and its performance was then measured with the behavior scale. Each session was composed of a theoretical explanation, a practical exercise, a final evaluation for the students about the given topic, and a time for questions and answers.

The exercise with fractions included a visual exercise, where the teacher or the robot would show a geometric figure divided by a certain number of colored parts and with blank spaces. After showing the image, the students were questioned about how many spaces were missing to complete the figure and then which fraction represented that portion.

For the sound propagation experiment, the teacher or the robot would work as a receiver, while the student would be the transmitter of a message. The idea was to say a phrase to the receiver at a certain distance and determine if the message was received. In the affirmative case, the student would increment the distance with the teacher or the robot and repeat the exercise until they fail to receive the phrase. When the instructor or robot was not capable of hearing, the student was asked to use a paper cone to transmit the message again and verify if, in that case, the message was transmitted successfully.

The exercise of metric system evaluation is shown in **Figure 4**. In this case, a two-meter band was used as a reference for putting the robotic platform at the beginning of the track. The students were required to use 20-cm bands to determine how many were required in order for the robot to move a particular distance.

The results obtained from these experiments demonstrated that the students showed a better learning process when using the robotic platform as a reference during the experiments than the case where the teaching staff was the only one interacting with them. Finally, a perception survey was also given to the students, where they expressed their interest in the activities and the use of a robot in class. In those questionnaires, the students showed signs of great interest. Also, during a final evaluation given to the students, it was observed how the ones that interacted with the robotic platform during the exercises achieved better scores than the ones that did not interact with the robot.

Figure 4.
Metric system evaluation with a robotic platform.

4.3 Attention span evaluation in high school students

High school students are more accustomed to the use of novel technological platforms and their use in daily life activities. For the same reason, the novelty of using robotic platforms during class may not be that interesting at first glance to them. The lack of novelty could even produce the reduction of the attention span during class. For this reason, it would be proven useful to evaluate the attention span of high school students during class, when the instructor is using a robotic platform, as illustrated in **Figure 5**.

The study described in [20] analyzes how much attention a group of high school students give during a mathematics class when the teacher uses a NAO robot.

Figure 5.
Use of a NAO robot in a high school class.

The methodology of the project was as follows: First, an electroencephalogram (EEG) equipment was used in some students during class, to obtain brain activity readings. Then, the teacher would give a pre-test to the group regarding the topic that would be given after. After the evaluation, the explanations and activities of the class were given by both the instructor and the robot, taking turns during the class to do so. The explanation procedure would continue after several classes until the topic was entirely given to the students. At the end of the last class, another evaluation was given to the students to compare their results with the first test.

The results of the group that interacted with the robot during the sessions were compared with another that received the same explanations and evaluations but without the NAO robot implementation. The EEG readings were also captured in the control group to generate the comparison as well.

While the class was given in both groups, a behavioral observation protocol was used to analyze the body behavior and attitudes of the students. Finally, at the end of the sessions, a little survey was given to every student and teaching staff that participated in the experimentation to analyze their perceptions about the use of a robotic platform during class.

The quantitative results came from filtering the EEG readings of the students to acquire the alpha and beta brain wave activity [21]. These two signals indicate the relaxation and concentration periods of a human being, allowing to determine in which moments the student was paying attention and to which parts of the class. The qualitative results came from the analysis of an Auzmendi attitude scale [22] that evaluates the anxiety, utility, motivation, and confidence the student shows during class.

By comparing both results, it was concluded that a robotic platform is capable of improving the attention span of students during class. However, it was also noticed that the attention was given most of the time to the robot's activities rather than the explanation, so a change in the class methodology and robot implementation is due.

4.3.1 Deployment considerations

The different situations described earlier clearly state that the use of SAR platforms in both healthcare and education improves the direct population that interacts with this technology. However, it is still needed to evaluate the feasibility of extending the use of these platforms nationwide.

According to the last survey of INEGI regarding the number of schools of different levels (from preschool to high school) [23], in 2018, there was a total of 244,117 schools around the country, from which 85.11% represents the public school system (see **Table 1** for the school distribution according to the educational level). Considering that public education covers 87.98% of Mexican students, it can be assumed that the financial cost of the government to implement these platforms would be extremely high. It would be necessary then to search for the collaboration between the government and enterprise to finance the implementation of these platforms in every single school.

Another point in consideration is the training of the personnel. Since the teaching staff is not accustomed to using these technologies, trained personnel need to instruct them about the correct use of the platforms, as well as how they can be implemented into their educational curricula effectively. In this case, the problem would be the lack of trained people, since the use of SAR platforms is relatively new in Mexico.

It should probably take around 12 years to develop a national implementation program that covers the creation of collaboration projects between the government

Educational level	Financial support	Number of schools	Number of students
Preschool	Public	73,574	4,139,977
	Private	16,005	751,025
Elementary	Public	87,756	12,678,241
	Private	9164	1,341,963
Secondary	Public	34,293	5,939,235
	Private	5396	597,026
High school	Public	12,138	4,237,524
	Private	5791	999,479

Table 1.
Number of schools in Mexico [23].

and enterprise, the acquisition of enough platforms to cover national population, the training of engineers for the platform manipulation and methodology design, and the training of the teaching staff of every school personnel.

The use of SAR platforms in the Mexican healthcare system is a little more complex to achieve. Since last year's change of administration, the healthcare system in Mexico has sustained multiple structural changes. It would then be necessary to wait at least 3 years until the new system is sufficiently accomplished to start considering the use of SAR technologies to support medical labor. At the same time, there are other problems regarding the number of medical professionals and medical access that made the suggestion of SAR platforms difficult. Nevertheless, those aspects are out of the scope of this chapter.

It can be said generally that the use of SAR technologies in Mexico is still in the phase of experimental evaluation, where these platforms are being used in particular situations to evaluate the future possibility of national implementation.

5. Conclusions

The different scenarios mentioned in this chapter show the importance that has been given to socially assistive robots in both occupational therapy sessions and educational purposes. The few examples show promising results, as well as the desire of researchers to continue evaluating how these platforms are capable of improving the quality of life of the population in different areas. However, it is still necessary to mention the obstacles that still need to be solved in the future.

The first problem is that currently, robotic platforms, as the ones mentioned through the chapter, are expensive, making it difficult for different national institutions to acquire one platform. At the same time, there are not enough people capable of operating and giving maintenance to the robot, which transforms the platform into obsolete after some time.

Another difficulty presented in this work is that every reference and experiment mentioned was only experimental. Hence, only a small population is capable of benefiting from the use of the platform. A socially assistive robot will only be utterly efficient until most of the population benefits from its use. The idea is to use these platforms to improve the quality of life of everyone. Unfortunately, it is still necessary to develop economic and social policies besides the technological that befits the use of such inventions.

Even though the use of robotic platforms is becoming more usual, there are still areas of opportunity that need to be considered before using this technology regularly. However, it does not mean that the advancements achieved so far are less critical. On the contrary, these studies demonstrate the feasibility of using robotics in a more human-like way, improving the way we communicate and perform our duties in society.

Author details

Edgar Lopez-Caudana[1*], Germán Eduardo Baltazar Reyes[1] and Pedro Ponce Cruz[2]

1 Tecnológico de Monterrey, Engineering and Sciences School, Ciudad de México, México

2 Writing Lab, TecLabs Tecnologico de Monterrey, Monterrey, Mexico

*Address all correspondence to: edlopez@tec.mx

References

[1] Feil-Seifer D, Mataric MJ. Defining socially assistive robotics. In: 9th International Conference on Rehabilitation Robotics, 2005. ICORR 2005; 2005. pp. 465-468

[2] Tapus A, Maja M, Scassellatti B. The grand challenges in socially assistive robotics. In: IEEE Robotics and Automation Magazine Special Issue on Grand Challenges in Robotics; 2007

[3] López N, Sánchez L. El aburrimiento en clases. Procesos psicológicos y sociales. 2010;**6**(1-2):1-43

[4] Santiago Benítez G, Caballero Álvarez R, Gómez Mayén D, Domínguez Cuevas A. El uso didáctico de las TIC en escuelas de educación básica en México. Revista Latinoamericana de Estudios Educativos (México). 2013;**43**(3): 99-131

[5] INEGI. La discapacidad en México, datos al 2014: versión 2017, INEGI; 2018

[6] La Razón Onlilne [Online]. 2017. Available from: https://www.razon.com.mx/salud/sabias-que-en-mexico-alrededor-del-6-de-la-poblacion-tiene-alguna-discapacidad/

[7] World Federation of Occupational Therapists, "wfot.org," World Federation of Occupational Therapists, 2012 [Online]. Available from: https://www.wfot.org/about-occupational-therapy

[8] World Federation of Occupational Therapists. Occupational Therapy and Assistive Technology, WFOT; 2019

[9] González Galván EJ, Martínez Delgado U, Abad Puentes J, Reyes Cortés F, Fernández Pérez DR. El uso de robots como apoyo en treapia ocupacional. Saberes y Ciencias, no. 38; 2015

[10] Secretaría de Educación Pública. Educación por Niveles [Online]. 2010. Available from: https://www.sep.gob.mx/es/sep1/educacion_por_niveles

[11] Subsecretaría de Educación Media Superior. Competencias Disciplinares Básicas organizadas en Campos Disciplinares (Componente de formación básica). In: Documento Base del Bachillerato General, México, Dirección General del Bachillerato; 2016. pp. 23-24

[12] Schleicher *A. PISA* 2018 Insights and Interpretations. Organisation for Economic Co-operation and Development; 2018

[13] Instituto Nacional para la Evaluación de la Educación. PLANEA. Resultados nacionales 2017; INEE; 2018

[14] Gutiérrez-Robledo LM, Ávila-Fematt FM, Montana-Álvarez M. La geriatría en México. El Residente. 2010;**5**(2):43-48

[15] Consejo Mexicano de Geriatría. Médicos Geriatras con Certificación Vigente; 2017

[16] Giraldo-Rodríguez L, Rosas-Carrasco O. Development and psychometric properties of the geriatric mistreatment scale. Geriatrics & Gerontology International. 2013;**13**(2):466-474

[17] Nava Lagunes AF, Pérez Zepeda MU, Lopez Caudana EO. Use of NAO robot in training of primary care clinicians for treatment of elderly patients. The Journal of Latin American Geriatric Medicine. 2017;**3**(2):53-56

[18] Instituto Nacional de Salud Pública. Encuesta Nacional de Salud y Nutrición; INEGI; 2018

[19] Ponce P, Molina A, Caudana EOL, Reyes GB, Parra NM. Improving education in developing countries using robotic platforms. International Journal on Interactive Design and Manufacturing (IJIDeM). 2019;**13**(4):1401-1422

[20] Lopez Caudana E, Baltazar Reyes G, Garcia Acevedo R, Ponce P, Mazon N, Membrillo Hernandez J. RoboTICs: Implementation of a robotic assistive platform in a mathematics high school class. In: 2019 IEEE 28th International Symposium on Industrial Electronics (ISIE), Vancouver; 2019

[21] Navarro R. Captación del EEG. Electroencefalografía; 2002

[22] Auzmendi Escribano E. Las actitudes hacia la matemática-estadística en las enseñanzas media y universitaria. Características y medición; 1992

[23] INEE. La Educación Obligatoria en México. Instituto Nacional para la Evaluación de la Educación; 2019

Section 2

New Applications in Industry 4.0

Chapter 4

Dynamic Compensation Framework to Improve the Autonomy of Industrial Robots

Shouren Huang, Yuji Yamakawa and Masatoshi Ishikawa

Abstract

It is challenging to realize the autonomy of industrial robots under external and internal uncertainties. A majority of industrial robots are supposed to be programmed by teaching-playback method, which is not able to handle with uncertain working conditions. Although many studies have been conducted to improve the autonomy of industrial robots by utilizing external sensors with model-based approaches as well as adaptive approaches, it is still difficult to obtain good performance. In this chapter, we present a dynamic compensation framework based on a coarse-to-fine strategy to improve the autonomy of industrial robots while at the same time keeping good accuracy under many uncertainties. The proposed framework for industrial robot is designed along with a general intelligence architecture that is aiming to address the big issues such as smart manufacturing, industrial 4.0.

Keywords: industrial robot, autonomy, intelligence architecture, high-speed sensing, coarse-to-fine strategy

1. Introduction

Performance of industrial robots in realizing fast and accurate manipulation is very important for manufacturing process, as it directly relates to productivity and quality. On the other hand, with manufacturing shifting from an old era of mass production to a new era of high-mix low-volume production, autonomous capability of industrial robots becomes more and more important to the manufacturing industry. Autonomy represents the ability of a system in reacting to changes and uncertainties on the fly.

Currently, off-line teaching-playback using a teaching pendant, or physically positioning a robot with a teaching arm, is supposed to be the main method for the applications of industrial robots. The method features a user-friendly interface developed by commercial robot manufacturers and is usually motion optimized and reliable so long as task conditions do not change. As detailed in [1], negative effects of nonlinear dynamics during high-speed motion may be pre-compensated in order to achieve accurate path tracking during the playback phase. However, it is impossible for a teaching-playback robot to adapt to significant variations in the initial pose of a working target or unexpected fluctuations during manipulation. CAD model-based teaching methods neither enable a robot to adapt to changes on the fly.

In [2], a view-based teaching-playback method was proposed to achieve robust manipulation against changes in task conditions with the use of artificial neural network (ANN). However, the approach is difficult for teaching jerky robot motion and cannot be applied to cases where high motion accuracy is required.

By utilizing external sensory feedback (e.g., vision), on-line control method may help a robot adjust to environmental uncertainty. Generally, accurate models with structured working environment are preconditions for implementation. However, in reality, accurate models are difficult to obtain. Regarding these issues, many adaptive approaches have been proposed (e.g., [3–7]) to address the control problem in the presence of uncertainty associated with a robot's kinematics model, mechanical dynamics, or with sensor-robot mapping. However, it is usually difficult to obtain satisfactory accuracy at a fast motion speed due to the complex dynamics and large mechanical inertia of a typical multi-joint industrial robot [8, 9].

In order to improve autonomy and to address on-line uncertainty attributed from a robot itself or external environment, ideally we need the feedback control of the robot in task space with much higher bandwidth than that of the accumulated uncertainty. Therefore, high-speed sensing and high-speed control based on high-speed feedback information should be realized. This kind of system has been developed decades ago such as the 1 ms sensor-motor fusion system as presented in [10]. However, from the viewpoint that easy integration with a commercial robot's black-box controller (or even to consider the compatibility with the Industrial 4.0 [11]) is also an important issue, there is still no practical framework that can effectively address these issues.

In this chapter, we present the dynamic compensation framework to improve the autonomy of industrial robots. The dynamic compensation concept [12–16] is implemented based on high-speed sensory feedback as well as a coarse-to-fine strategy inherited from the macro–micro method [17, 18]. It should be noted that the macro–micro concept had been proposed several decades ago with the aim of enhancing system bandwidth for rigid manipulators and suppressing bending vibrations for flexible manipulators and this is not the scope of this study. In order to show the effectiveness of the proposed framework, several application scenarios are also presented.

2. Methodology

In this section, dynamic compensation framework under a hierarchical intelligent architecture would be presented. For the issue of asymptotic convergence, an intuitive analysis based on a simplified model will then be introduced. System integration with an industrial robot under the proposed framework would be addressed.

2.1 Dynamic compensation framework

The proposed framework for improving autonomy of industrial robots is based on a coarse-to-fine strategy as shown in **Figure 1**. Intelligence of a system is considered to be made up of two parts: action-level intelligence for motion control and planning-level intelligence for motion and task planning. Action-level intelligence represents low-level layer of the intelligence architecture and is referred to adaptability for motion control without sacrificing motion speed and absolute accuracy simultaneously, and we consider it as the foundation in implementing high-level intelligence. Real-time adaptation to both system uncertainty and environmental uncertainty enables a robot to focus on implementing high-level intelligence.

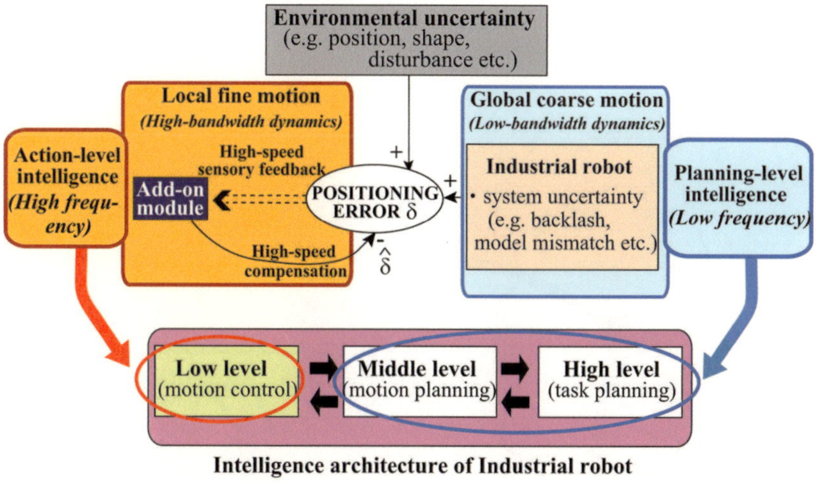

Figure 1.
Proposed framework.

Ideally, action-level intelligence is realized with high-frequency update rate to address the high-frequency part of on-line uncertainties and changes, whereas planning-level intelligence is allowed to implement with low-frequency update rate to tackle with the low-frequency part of uncertainties and changes. For implementation, traditional industrial robots are designated to conduct coarse global motion by focusing on planning-level intelligence. Concurrently, an add-on robotic module with high-speed actuators and high-speed sensory feedback is controlled to realize fine local motion with the role of implementing action-level intelligence.

2.2 Analysis on asymptotic convergence with a simplified model

In order to grasp the basic idea of the proposed framework, intuitive analysis on asymptotic convergence with a simplified model is provided with the assumption that the entire system is regulated in image space [13, 14].

As shown in **Figure 2**, an arbitrary industrial robot is controlled from position C toward position A (for simplification, the target is assumed to be motionless) with visual feedback information from a global camera. A direct-driven add-on module with high-speed actuators and high-speed sensory feedback is configured to its end effector. Specifically, we take the example of high-speed vision sensing for the add-on module. Initially, tool point B is assumed to overlap with tool point C. We refer to image features of target A and those of the robot's tool as $\boldsymbol{\xi}_a$ and $\boldsymbol{\xi}_c$, respectively, from visual feedback of the global camera, and error $\boldsymbol{e}$ for regulation is noted as

$$\boldsymbol{e} = \boldsymbol{\xi}_c - \boldsymbol{\xi}_a \tag{1}$$

Noting that $\dot{\boldsymbol{\xi}}_c$ can be divided into two parts, motion effects corresponding to the main robot and the compensation module,

$$\dot{\boldsymbol{\xi}}_c = \boldsymbol{J}_r \dot{\boldsymbol{\theta}}_m + \boldsymbol{J}_c \dot{\boldsymbol{\theta}}_c \tag{2}$$

where $\dot{\boldsymbol{\theta}}_m$, $\dot{\boldsymbol{\theta}}_c$ represent joint velocity vectors of the main robot and the compensation module, respectively, and $\boldsymbol{J}_r$ and $\boldsymbol{J}_c$ are the Jacobians (mapping from joint space to image space) of the main robot and the compensation module, respectively. We assume that the compensation module is not activated ($\dot{\boldsymbol{\theta}}_c = 0$) and stays

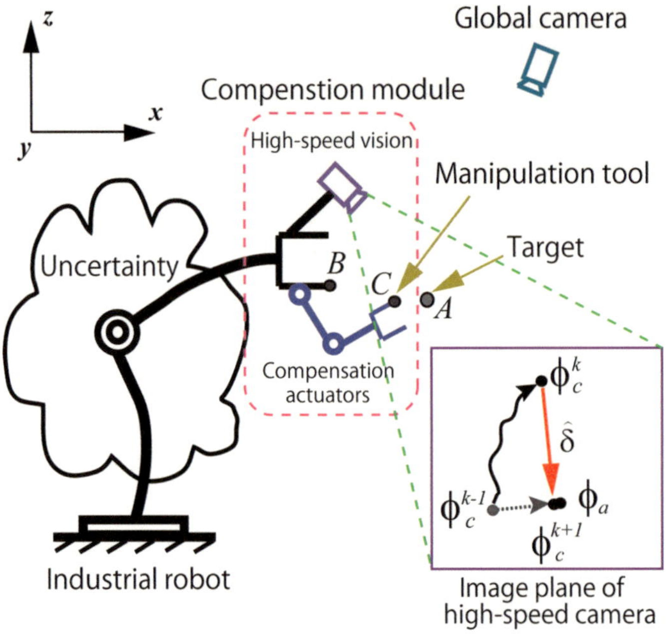

Figure 2.
A simplified model for addressing the dynamic compensation concept. Tool point C is controlled to align with target position A with the help of a compensation module, although the position of B is not certain due to the systematic uncertainty of the main robot. Image features of the global camera is represented by $\boldsymbol{\xi}$, and image features of the high-speed camera is represented by $\boldsymbol{\phi}$.

still to retain the overlap between B and C. With the ideal case, exponential convergence of error regulation (for instance, $\boldsymbol{\xi}_c$ converges to $\boldsymbol{\xi}_a$) can be obtained if we apply feedback control, such as

$$\dot{\boldsymbol{\theta}}_m = -\boldsymbol{\omega} \boldsymbol{J}_r^+ \boldsymbol{e} \tag{3}$$

where $\boldsymbol{\omega}$ is a constant positive-definite coefficient matrix and $\boldsymbol{J}_r^+$ represents the pseudo-inverse of $\boldsymbol{J}_r$. However, in practice, the ideal visual-motor model of $\boldsymbol{J}_r^+$ for an industrial robot is not available and is usually estimated with errors due to systematic uncertainties and inaccurate camera calibration. We denote the uncertain part as $\Delta \boldsymbol{J}_r^+$. The error dynamics with feedback control then become

$$\dot{\boldsymbol{\xi}}_c = -\omega \boldsymbol{J}_r (\boldsymbol{J}_r^+ + \Delta \boldsymbol{J}_r^+) \boldsymbol{e} \tag{4}$$

and

$$\dot{\boldsymbol{e}} = -\omega \boldsymbol{e} - \boldsymbol{\delta} \tag{5}$$

where δ represents the projected uncertainty in image space of the global camera. In this case, for instance, $\boldsymbol{\phi}_c^{k-1}$ moves to $\boldsymbol{\phi}_c^k$ rather than $\boldsymbol{\phi}_a$ in image space of the high-speed vision system due to uncertainty, as shown in **Figure 2**. It should be pointed out that in spite of the uncertain term $\boldsymbol{\delta}$, the system is still assumed to conduct coarse positioning in the direction of the neighborhood of the target with the visual feedback of the global camera.

Now, let the compensation module be activated with motion such that $\boldsymbol{J}_c \dot{\boldsymbol{\theta}}_c = \dot{\boldsymbol{\phi}}_c$. Let $\hat{\boldsymbol{\delta}}$ represents the compensation module's motion observed by the global camera, and let γ represents the conversion factor such that $\dot{\boldsymbol{\phi}}_c = \gamma\dot{\boldsymbol{\xi}}_c$. Note that γ can be perceived as time-invariant as the lens systems of both cameras are fixed and the relative position between the two cameras can be assumed as constant for a short-time period. The closed-loop system then becomes

$$\begin{aligned}\dot{\boldsymbol{e}} &= -\omega\boldsymbol{e} + \hat{\boldsymbol{\delta}} - \boldsymbol{\delta} \\ &= -\omega\boldsymbol{e} - \tilde{\boldsymbol{\delta}}\end{aligned} \tag{6}$$

where $\tilde{\boldsymbol{\delta}} = \boldsymbol{\delta} - \hat{\boldsymbol{\delta}}$. In order to obtain the update law for $\hat{\boldsymbol{\delta}}$, we choose the following Lyapunov function candidate

$$V(\boldsymbol{e}, \boldsymbol{\delta}) = \mathbf{e}^T\boldsymbol{P}\boldsymbol{e} + \tilde{\boldsymbol{\delta}}^T\boldsymbol{\Gamma}^{-1}\tilde{\boldsymbol{\delta}} \tag{7}$$

where $\boldsymbol{P}$ and $\boldsymbol{\Gamma}$ are two symmetric positive-definite matrices. Suppose that the direct-driven compensation module is feedback controlled by 1000 Hz of high-speed vision and the uncertain term $\boldsymbol{\delta}$ due to the main robot which has big inertia can be approximated as a constant unknown term during the 1 ms feedback control cycle. Therefore, $\dot{\tilde{\boldsymbol{\delta}}} = -\dot{\hat{\boldsymbol{\delta}}}$, and the time derivative of V is given by

$$\begin{aligned}\dot{V} &= \dot{\boldsymbol{e}}^T\boldsymbol{P}\boldsymbol{e} + \boldsymbol{e}^T\boldsymbol{P}\dot{\boldsymbol{e}} - 2\tilde{\boldsymbol{\delta}}^T\boldsymbol{\Gamma}^{-1}\dot{\hat{\boldsymbol{\delta}}} \\ &= -2\omega\boldsymbol{e}^T\boldsymbol{P}\boldsymbol{e} - 2\tilde{\boldsymbol{\delta}}^T\boldsymbol{\Gamma}^{-1}\left(\boldsymbol{\Gamma}\boldsymbol{P}\boldsymbol{e} + \dot{\hat{\boldsymbol{\delta}}}\right)\end{aligned} \tag{8}$$

Apparently, $\dot{V} = -2\omega\boldsymbol{e}^T\boldsymbol{P}\boldsymbol{e} \le 0$ if we choose the update law for $\hat{\boldsymbol{\delta}}$ as

$$\dot{\hat{\boldsymbol{\delta}}} = -\boldsymbol{\Gamma}\boldsymbol{P}\boldsymbol{e} \tag{9}$$

And resultantly, the control law for the compensation module should satisfy the following condition:

$$\boldsymbol{J}_c\dot{\boldsymbol{\theta}}_c = -\gamma^{-1}\boldsymbol{\Gamma}\boldsymbol{P}\boldsymbol{e} \tag{10}$$

From the analysis above, we claim that asymptotic convergence is achievable using the proposed dynamic compensation in spite of systematic uncertainty in the main robot. Several issues should be noted here. First, the same conclusion can be drawn no matter how the main robot is controlled (in task space, as here, or joint space), and compensation capability can be further enhanced due to the fact that the control frequency of most commercial industrial robots is smaller than the 1000 Hz feedback control of the directly driven compensation module. Second, although dynamics are not fully incorporated within the analysis, our claim is still reasonable under the condition that the compensation actuator has a different bandwidth from that of the main robot. Third, although we have assumed the target to be motionless above, it is reasonable to apply the same analysis to cases where the target is moving but its motion is negligible in the context of 1000 Hz high-speed vision sensing. Moreover, although several robust and adaptive control approaches have been proposed (e.g., [3]) for direct control of robots with uncertain kinematics and dynamics, we note the advantages of our method in following two aspects:

1. The method here decouples the direct-driven compensation module and the main industrial robot and requires no changes to the main robot's controller.

On the contrary, traditional adaptive control methods need to directly assess the inner loop of a robot's controller (mostly not open), which is usually considered difficult both technically and practically.

2. It is difficult for traditional adaptive control methods to realize high-speed and accurate adaptive regulation due to the main robot's large inertia and complex nonlinear dynamics. With the philosophy of motion decoupling as well as adopting high-speed vision to sense the accumulated uncertainties, the proposed method here enables a poor-accuracy industrial robot to realize high-speed and accurate position regulation by incorporating a ready-to-use add-on module.

To summarize, the proposed dynamic compensation involves three important features:

1. The compensation module should be controlled accurately and sufficiently fast. Ideally, it has a much larger bandwidth than that of the main robot.

2. The sensory feedback for the compensation module should be sufficiently fast in order to satisfy the assumption $\dot{\tilde{\boldsymbol{\delta}}} = -\dot{\hat{\boldsymbol{\delta}}}$.

3. The error value $\boldsymbol{e}$ is the relative information between the robot's tool point and the target in image coordinates, which can be observed directly.

Finally, it should be noted that since the add-on module works independently of the main robot's controller, optimal control of the system is another issue to address and is beyond the scope of this chapter.

2.3 Integration with conventional industrial robots

The proposed framework can be easily integrated with existing industrial robots. Usually, inner control loops of conventional industrial robots are black-box to end users due to issues concerning safety and intellectual property, and limited functions such as trajectory planning are available to users through interfaces provided by robot makers. In other words, it is difficult to incorporate external sensory information into the robot's inner control loop for motion control. On the other hand, as shown in **Figure 3**, control scheme for an arbitrary industrial robot and the add-on compensation module is separated with the proposed framework. Therefore, the compatibility of the proposed method is good as the industrial robot itself can be perceived as a black-box and users only need common interfaces for integration.

In the dynamic compensation framework, an industrial robot is designated for fast and coarse motion. Therefore, efforts for trajectory planning can be greatly reduced compared to traditional teaching-playback methods as well as adaptive control methods based on external sensory feedback. Coarse motion planning of the industrial robot can be either in a semiautonomous way or in an autonomous way. In the former method, teaching points covering a target trajectory can be very sparse as long as the target trajectory can be accessed by the add-on module. In the later method, coarse trajectory can be planned by utilizing external sensors (e.g., camera), and the calibration for sensor-motor mapping can be rough and easy. However, for conventional methods utilizing external sensors to realize accurate motion control of the industrial robot, calibration for sensor-motor mapping can be very complex and difficult.

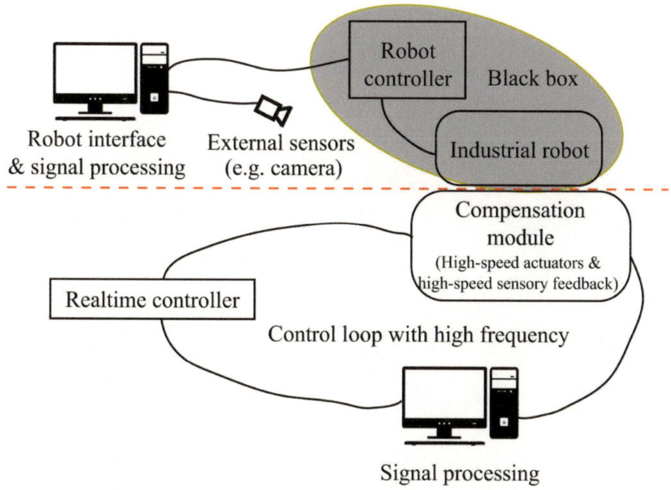

Figure 3.
Integration with conventional industrial robots.

Such as many other traditional methods, implementation of motion planning based on external sensory feedback involves two aspects: calibration for the mapping between the coordinates of sensory information and those of an industrial robot and signal processing to extract key points for motion planning from the sensory space according to a specified task. From the perspective of motion control, application tasks of industrial robots can be categorized into two kinds: set point motion control and trajectory motion control. For conventional methods, extracting key points and then realizing point-to-point motion control with accurate sensor-motor models for tasks involving set point motion (e.g., peg-in-hole task) is relatively easy to assure good accuracy. However, for tasks involving continuous trajectory control based on extracted key points and accurate sensor-motor models, good accuracy is still hard to realize due to the complex dynamics of an industrial robot during on-line moving. On the other hand, since an industrial robot is only asked to realize coarse motion in the proposed framework, calibration can be rough and easy, and motion errors due to the complex dynamics or even mismatched kinematics can be allowable, as long as the accumulated error is within the work range of the add-on compensation module. Application tasks with both set point motion control and trajectory motion control will be implemented with the proposed framework. Firstly, simplified peg-and-hole alignment in one dimension will be presented in Section 3. And then, contour tracing in two dimensions will be introduced in Section 4.

3. Application scenario 1: peg-and-hole alignment in one dimension

The simplified peg-and-hole alignment task was conducted to test the efficiency of the proposed method in realizing fast and accurate set point regulation under internal and external uncertainties [13]. The experimental testbed is shown in **Figure 4**. The add-on compensation module had one degree of freedom (DOF). A workpiece (metal plate with six randomly configured holes) was blindly placed on a desk for each experiment trial. The holes were 2 mm along the x-direction and were elongated in the y-direction to account for the fact that compensation was carried out only in the x-direction. A mechanical pencil with a diameter of 1.0 mm acting as the peg was attached to the linear compensation actuator, and the insertion

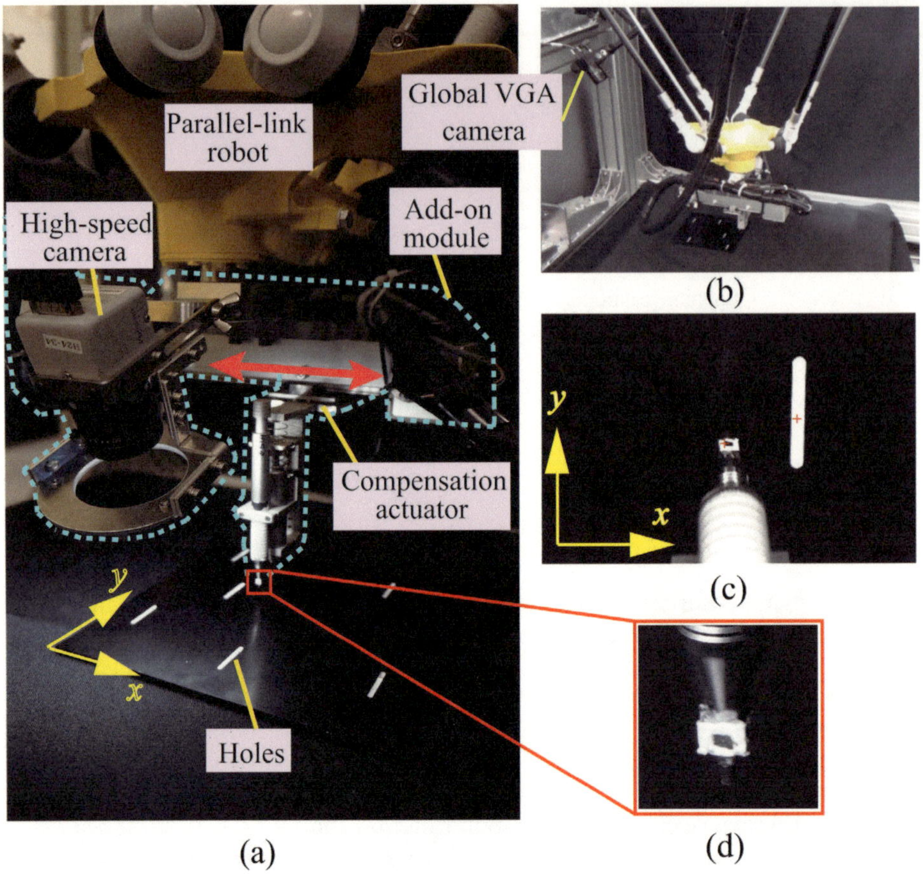

Figure 4.
Experimental system [13]. (a) Overall setup: A one-DOF (x-direction) add-on module and a commercial parallel-link robot. (b) Global VGA camera for coarse motion planning of the parallel-link robot. (c) Detected marker and reference position (center of the nearest hole). (d) Marker representing the peg's position.

action was driven by an on–off solenoid. The insertion action was activated only if the error between the peg and the center of the hole in the x-direction was smaller than 0.8 pixels (corresponding to 0.112 mm) and lasted for more than 0.02 s. The insertion lasted for 0.3 s. We sought to insert the peg at the center of these holes. As can be seen from **Figure 4**, the holes formed the white parts of the otherwise black workpiece. Section 3.3.2 describes the process of detecting and obtaining the positions of these holes.

3.1 High-speed vision system with traditional configuration

As analyzed in Section 2.2, feedback information in task space for the add-on module should be high speed in order to satisfy the assumption $\dot{\tilde{\boldsymbol{\delta}}} = -\dot{\hat{\boldsymbol{\delta}}}$. Here, high-speed vision system with traditional configuration (namely, imaging and image processing are conducted separately as shown in **Figure 5(a)**) was introduced. A Photron IDP-Express R2000 high-speed camera [19] (made by Photron, Japan) was used with the eye-in-hand configuration. The camera is capable of acquiring 8-bit monochrome or 24-bit color frames with the resolution 512 × 512 pixels at a frame rate of 2000 fps. It was connected to an image processing PC (OS, Windows 7 Professional; CPU, 24 core, 2.3 GHz Intel Xeon; Memory, 32 GB; GPU, NVIDIA

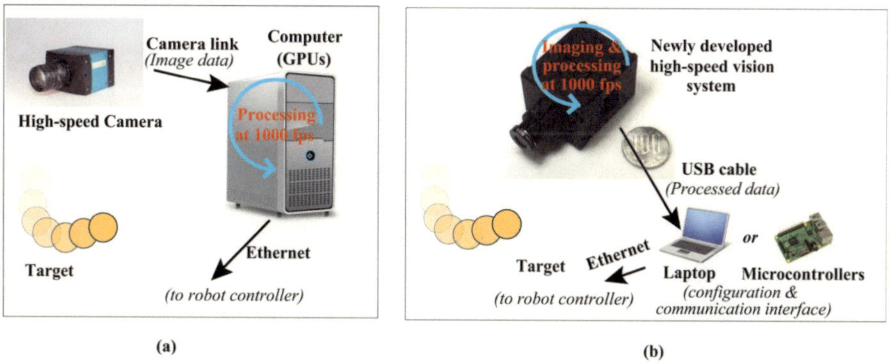

Figure 5.
High-speed vision system [25]. (a) Traditional system with imaging and image processing conducted separately. (b) New high-speed vision system with high-speed imaging and processing integrated in one chip.

Stroke	Maximum velocity	Maximum acceleration	Weight
100 mm	1.6 m/s	200 m/s^2	0.86 kg

Table 1.
Spec. of actuator for the one-DOF add-on module prototype.

Quadro K5200) (made by Dell Inc., USA), and the high-speed camera was configured with a working frame rate of 1000 fps.

Since control was limited to one dimension (along the x-direction in our case), the peg and the holes only needed to be aligned along the x-axis in the images. The peg was tracked using a marker fixed on the mechanical pencil at some distance from its tip (**Figure 3(b)**). In the captured images, it corresponded to roughly 9×9 pixel patches, and we employed a simple template-based search to find the location of the marker by minimizing the mean squared error. Following marker identification, we calculated its location at sub-pixel accuracy by computing image moments on the center patch. After locating the peg, we searched for the hole on a row in the image at a fixed distance from the detected marker in the y-direction. This image row was effectively binarized apart from the edge regions around the holes. We therefore searched for consecutive white regions (holes) and selected the one with center closest to the peg in the x-direction. The hole position was also computed with sub-pixel accuracy using image moments over the non-black region on the searched row. The processing ran within a millisecond using CUDA [20] to enable 1000 fps tracking of the positions of the pen and the hole. The high-speed camera is configured in such a way that the peg and each hole were both visible, and the relative error in image coordinates was sent to the real-time controller (**Figure 3**) by an Ethernet at a frequency of 1000 Hz. Since the high-speed camera was configured as the eye-in-hand, uncertainties due to the main robot as well as the external environment can be resultantly perceived as the variations of the hole's position, and they are accumulated within the relative error toward the peg.

3.2 Add-on compensation module with one DOF

The one-DOF add-on module with linear actuation was developed with specifications presented in **Table 1**. In accordance with the proposed dynamic compensation framework, the module was designed with large acceleration capability as well as being lightweight. The high-speed camera addressed above was configured with a field of view of approximately 70 mm within the motion range of the linear actuator. Therefore, we had an approximate conversion of 1 pixel to be 0.14 mm.

As indicated in Section 2.2, control law of the compensation module should be developed according to Eq. (10). Obviously, Eq. (10) can be held with simple proportional derivative (PD) control or some other methods such as precompensation fuzzy logic control (PFLC) algorithm [21].

3.3 Coarse motion planning of industrial robot

A four-DOF parallel-link robot capable of high-speed motion was deployed to execute coarse global motion. A low-cost Video Graphics Array (VGA) camera (made by SONY, Japan) was mounted on the frame of the system and directed at the workspace. Using this camera, we fully automate the teaching task by detecting the rough position of holes in the main robot's coordinates.

3.3.1 Calibration issue

With the proposed framework, an exact calibration was unnecessary. Therefore, we did not worry about the intrinsic calibration of the VGA camera. Calibration simply involved computing the planar homography between the workpiece (a metal plate) in the main robot's coordinates and the image plane of the VGA camera. This was done by letting the main robot move to four points $\boldsymbol{X}_i = \{x_i, y_i, k_i\}^T$ directly above the workpiece and marking the corresponding locations $\boldsymbol{X}_{i'} = \{x_{i'}, y_{i'}, k_{i'}\}^T$ in the image. Here, the points $\boldsymbol{X}_i$ and $\boldsymbol{X}_{i'}$ are homogeneous coordinates representing the 2D coordinates $\{x_i/k_i, y_i/k_i\}^T$ and $\{x_{i'}/k_{i'}, y_{i'}/k_{i'}\}^T$. The four points could be chosen randomly such that no three points were collinear. Using these four point correspondences $\boldsymbol{X}_i \leftrightarrow \boldsymbol{X}_{i'}$, we computed the homography $\boldsymbol{H} \in \Re^{3\times3}$ [22] by

$$\boldsymbol{X}_{i'} = \boldsymbol{H} \cdot \boldsymbol{X}_i (i = 1, \dots, 4). \tag{11}$$

Since we only needed a coarse homography, the procedure of choosing four point correspondences was rough and easy to implement. While we only consider one-dimensional compensation in this task, the same calibration procedure applies to full 3D compensation over a limited depth range so long as holes on the workpiece are observable by the high-speed camera for fine compensation. Hence, it would not be necessary to obtain 3D measurements using a stereo camera configuration or some similar mechanism. Since the camera was fixed in relation to the main robot's coordinate frame, the calibration procedure was implemented once and only needed to be performed again if the camera was moved or if the height of the workspace changed drastically.

3.3.2 Hole detection and motion planning

Usually, the model of holes on the workpiece should be known in order to detect them and calculate their locations. Here, we simplified the detection problem by utilizing the fact that the holes formed the white area of the black workpiece. The holes were identified and their locations were computed using image moments. The resulting points were transferred to the main robot's coordinate system using the homography computed in Section 3.3.1. Since the calibration mapping between the image and the robot's coordinates was rough, the detected points expressed in the robot's coordinates should reside in the neighbor area of their corresponding hole.

Following this, the shortest path connecting all of these points was calculated as a traveling salesman problem (TSP) [23]. Finally, the route (all points in order) was sent to the controller of the parallel-link robot to generate the corresponding motion, with the maximum motion speed set at 2000 mm/s.

3.4 Experimental result

The result of one experimental trial for continuous peg-and-hole alignment for six holes in the workpiece is shown in **Figure 6(a)**. The workpiece was placed randomly. **Figure 6(b)** shows the details of the second alignment. It can be seen that while the parallel-link robot executed coarse positioning at a high speed (maximum speed: 2000 mm/s), the hole's image position from high-speed vision in the x-direction did not become stable until after 2.1 s due to the fact that its rotational axis exhibited significant backlash. Nevertheless, the compensation module realized

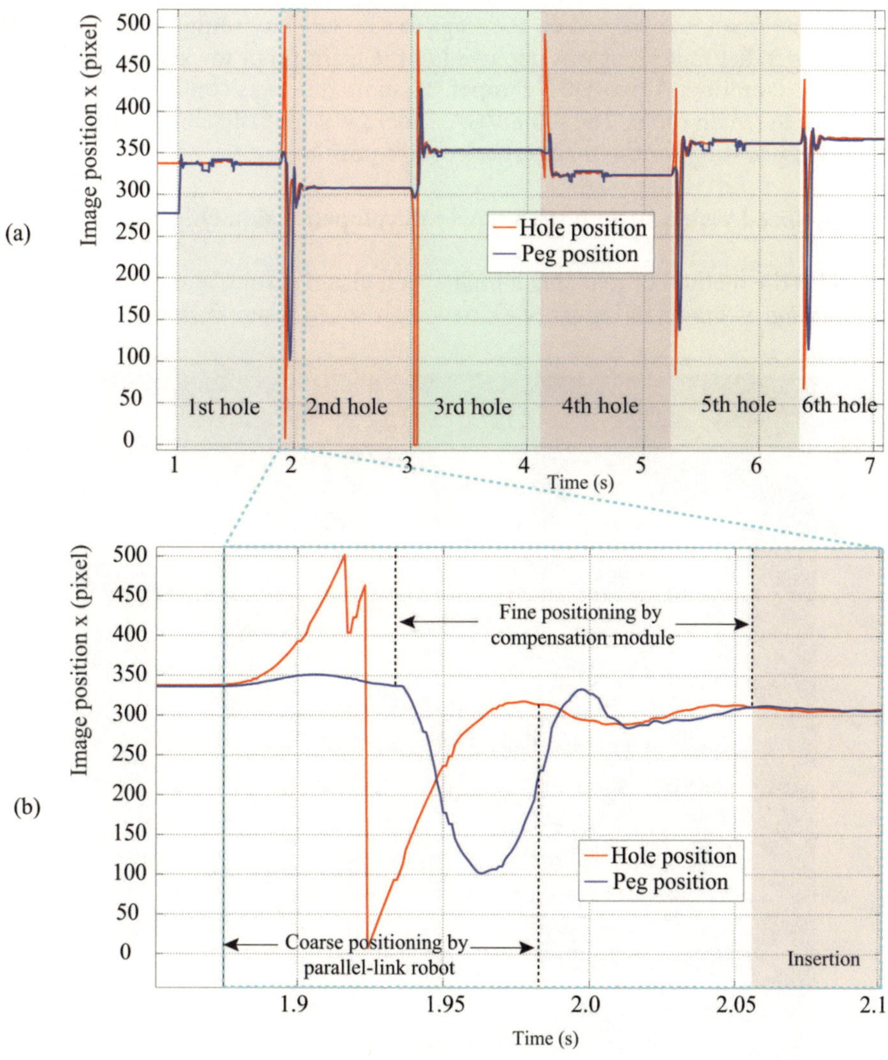

Figure 6.
Image feature profiles of pegand-hole alignment process [13]. (a) One experimental trial of continuous peg-and-hole alignment for six holes. (b) Zoomed details of the second alignment.

fine alignment within 0.2 s within an accuracy of 0.1 mm. For 20 trials with different positions of the workpiece, all alignments were satisfactory as the proper insertions were observed [13]. A video for the peg-and-hole alignment task can be found on the website [24].

4. Application scenario 2: contour tracing in two dimensions

The contour tracing task was conducted to verify the proposed method in realizing fast and accurate trajectory motion control under internal and external uncertainties [25]. Robotic contour tracing (contour following) is a useful technique in manufacturing tasks such as welding and sealing. Fast and accurate contour tracing under uncertainty from both a robot system and external environment is quite challenging. In this study, the task was confined in two dimensions. The experimental testbed is shown in **Figure 7**. The target is a closed curve with irregular contour pattern (2.5 mm width) printed on a paper which is placed on a stage. External disturbance is exerted on the stage to simulate environmental uncertainty. The same parallel-link robot was deployed as the main robot to execute the fast, coarse global motion. A two-DOF compensation module was configured at the end effector of the main robot. The same VGA camera was globally configured to realize the motion planning for the main robot's coarse positioning.

4.1 High-speed vision system with newly developed vision chip

Besides the method of realizing a high-speed vision system with traditional configuration as shown in **Figure 5(a)**, new high-speed vision system (**Figure 5(b)**)

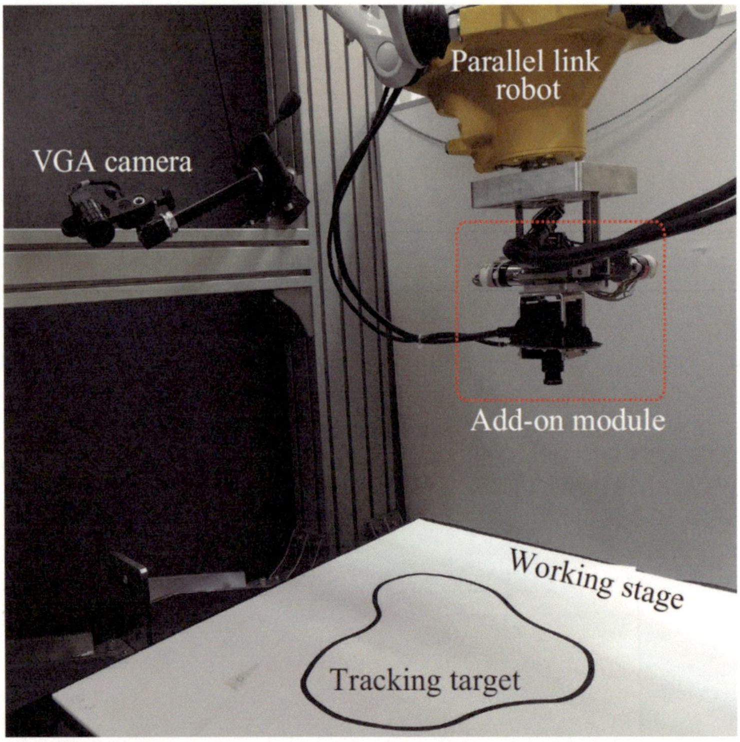

Figure 7.
Experimental setup for contour tracing task.

with newly developed vision chip is introduced. The new vision chip combines high-frame-rate imaging and highly parallel signal processing with high-resolution, high-sensitivity, low-power consumption [26]. The 1/3.2-inch 1.27 Mpixel 500 fps (0.31 Mpixel 1000 fps 2 × 2 binning) vision chip is fabricated with 3D-stacked column-parallel analog-to-digital converters (ADCs) and 140 giga-operations per second (GOPS) programmable single instruction multiple data (SIMD) column-parallel processing elements (PEs) for high-speed spatiotemporal image processing. The programmable PE can implement high-speed spatiotemporal filtering and enables imaging and various image processing such as target detection, recognition, and tracking on one chip. By realizing image processing on the chip, it can suppress power consumption to maximum 363 mW at 1000 fps. Comparing with conventional high-speed vision system, the new high-speed vision system will greatly save space and energy and is very suitable for compact usage in robotic applications. The high-speed vision was configured to work at 1000 fps with a resolution of 648 × 484. Overall latency of high-speed visual feedback was measured to be within 3.0 ms [25].

4.2 Add-on compensation module with two DOFs

In order to accompany with the parallel-link robot to realize the two-dimensional contour tracing task, an add-on module prototype capable of realizing fine compensation in two dimensions was developed. The add-on module was with two orthogonal linear joints, and specifications for the actuators of the module were estimated by an accelerometer and are shown in **Table 2**. The total weight of the module was about 0.27 kg. The high-speed vision was configured on the moving table of the add-on module, and the tracing task was implemented in such a manner that the high-speed vision was guided to travel along the curve with the curve's center accurately aligned with the center (324,242) of the high-speed vision's images.

4.3 Coarse motion planning of industrial robot

The same as the peg-and-hole alignment task, the main robot's motion was planned using vision information from the globally configured VGA camera. The implementation involved exactly the same with the last task: a rough calibration and image processing to extract key points (via-points) of a target contour path considering the limited working range of the add-on module. Extraction of key points of a target contour was implemented in the following manner [25]:

1. As shown in **Figure 8(a)**, the image was binarized with a proper threshold, and a start point p_0 on the target contour was determined with the nearest distance to a user predefined point. p_c was initialized as p_0.

2. A probing circle with its center at p_c was used to detect the intersection p_d with the target contour by predefined extraction direction.

Joint	Stroke	Max. velocity	Max. acceleration
x	20 mm	600 mm/s	63 m/s^2
y	20 mm	650 mm/s	70 m/s^2

Table 2.
Spec. of actuators for the two-DOF add-on module prototype.

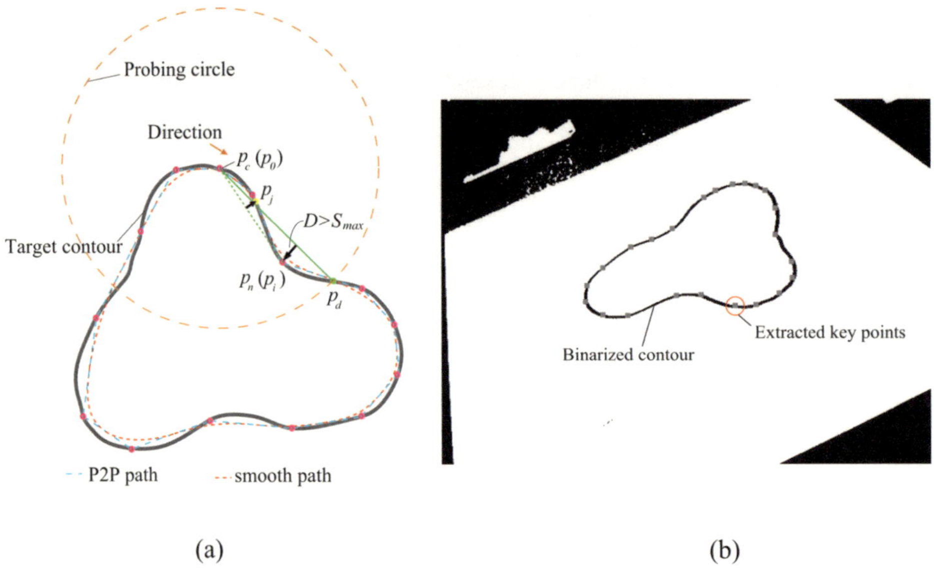

Figure 8.
Coarse motion of the main robot. (a) Method for extraction of key points. (b) Extracted key points.

3. Point p_i starting from p_c along the extraction direction with a small step size δ was examined. If its distance to chord $\overrightarrow{p_c p_d}$ was within the work range (S_{max}) of the compensation module in image space, we then continue to examine the next point by increasing one step further. Otherwise, p_i was elected as the new extraction point p_n. With insertion of the new point p_n, points between p_c and p_n should be re-examined to see whether distance from p_j to chord $\overrightarrow{p_c p_n}$ was bigger than S_{max} or not. If it was true, another new extraction point at p_j should be inserted, and a recursive check for points between p_c and p_j should be conducted. Until all points between p_c and p_n were secured (the distance from each point to the corresponding chord was smaller than S_{max}), move the probing circle by updating p_c with p_n. And then the algorithm will return back to the last step until all the discretization points of the target contour were visited. The distance from p_i to chord $\overrightarrow{p_c p_d}$ was represented as D and was calculated by

$$D = \frac{|\overrightarrow{p_c p_i} \times \overrightarrow{p_c p_d}|}{|\overrightarrow{p_c p_d}|} \tag{12}$$

4. Points $[p_0, p_1, \ldots, p_n]$ were the key points extracted from the target contour.

Usually, a commercial robot controller enables different methods of on-line path generation with selected key points. As an example, as shown in **Figure 8(a)**, a point-to-point (P2P) method that generated a path strictly passing through all key points with nonconstant velocity was included. On the other hand, a smooth path (100% smoothing factor) method would achieve a constant velocity profile while at the same time the exact generated trajectory would be not known to the user in advance. In many industrial applications, contour tracing with constant speed has significant advantages. It not only achieves good energy efficiency by reducing unnecessary acceleration and deceleration but also obtains better working performance in cases where work timing is critical, such as in welding. In this task, we intended to perform contour tracing with a constant speed of the main robot. Therefore the smooth path (100%) method was adopted to control the main robot. However, this introduced additional source of uncertainty to the main robot's trajectory. One case of the extracted key points is shown in **Figure 8(b)**.

4.4 Experimental result

Before each experiment, the paper with irregular contour pattern (2.5 mm width) printed on it was placed randomly on the working stage. Key points (shown in **Figure 8(b)**) for coarse tracing motion of the parallel-link robot were extracted. The parallel-link robot was set to be 100% smooth motion with 400 mm/s speed. Along with the coarse tracing by the parallel-link robot, the compensation module was realizing fine local compensation in two dimensions with the visual feedback information from the new high-speed vision system.

(a)

(b)

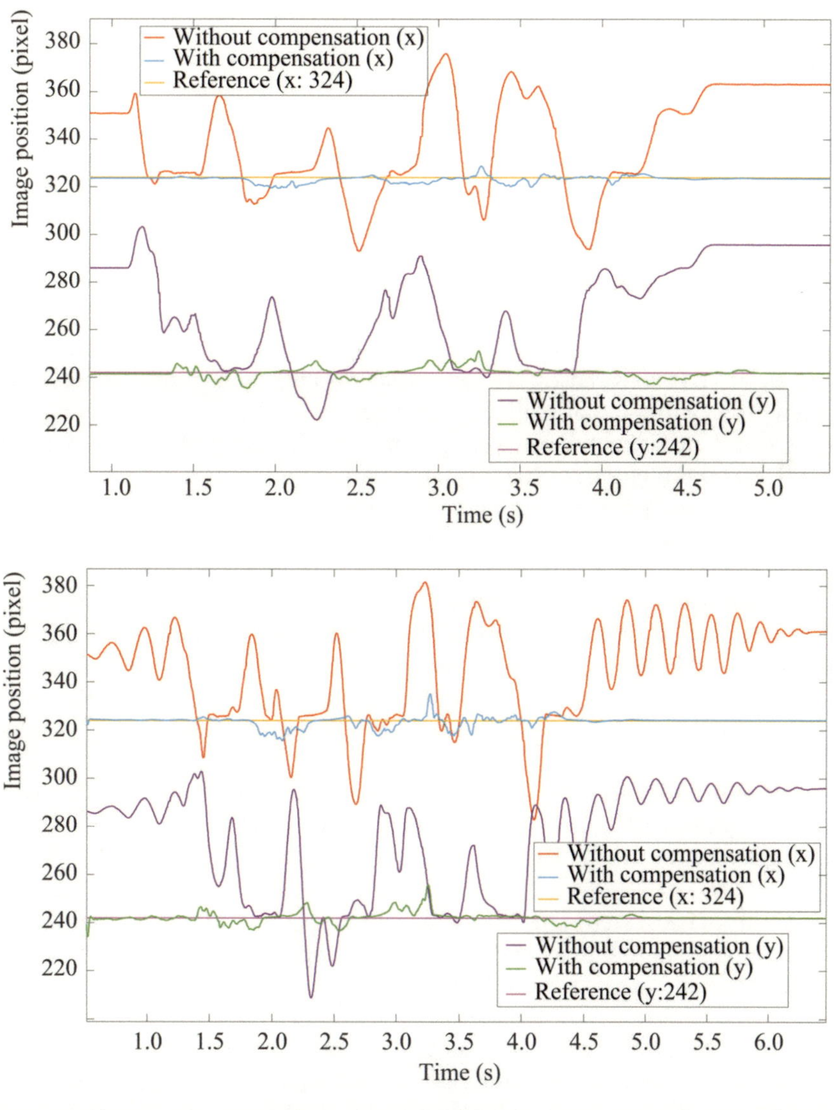

Figure 9.
Result of contour tracing [25]. (a) Image profile of contour tracing without external disturbance. (b) Image profile of contour tracing with random external disturbance.

Tracing result of one trial without external disturbance is shown in **Figure 9(a)**. It was obvious that with motion planning by the global visual information, only coarse tracing by the parallel link was realized. With the cooperation between the parallel-link robot's global motion and the add-on module's local compensation, fine tracing with maximum error around 6 pixels (corresponding to 0.288 mm in the experimental setup) was realized [25].

In order to simulate external disturbance to the working target, the stage was manually vibrated in a random manner during the tracing process. The image profile of one trial is shown in **Figure 9(b)**. It demonstrated that the proposed dynamic compensation robot was capable of realizing accurate tracing even under disturbance from external environment, with the maximum tracing error around 9 pixels (corresponding to 0.432 mm) [25]. A video for the contour tracing task can be found on the website [27].

5. Discussion

5.1 For wider applications

The two application scenarios addressed above held the same fact that the add-on modules were directly utilized in task implementation and the payloads for the add-on module were small. As addressed in the analysis of the proposed framework, an add-on compensation module was assumed to have high-speed motion capability and ideally have a much larger bandwidth than that of the main robot. Therefore, payload for the add-on module should be very limited if the add-on module is directly utilized in task implementation. Then the readers may wonder whether the proposed framework is adoptable for wider applications where payloads for robots could be large. In these cases, we can actually utilize the dynamic compensation framework as a method of realizing autonomous teaching (programming) for an industrial robot. Let us take an example of a certain task with trajectory motion control in two dimensions where a manipulation tool (e.g., a torch for arc welding) would be too heavy for the add-on compensation module. The autonomous teaching procedure may implement as follows:

1. **Learning step**. The industrial robot with an add-on compensation module configured at its end effector realizes autonomous contour tracing (as addressed in Section 4) for a specific contour target (e.g., a seam for welding task). Configurations (in configuration space or joint space) of the industrial robot and the add-on module are memorized during the tracing process.

2. **Programming step**. Integrating the data of the add-on module into the configuration of industrial robot at each time point to generate the synthesized trajectory for the industrial robot.

3. **Manipulation step**. The industrial robot replaces the add-on module with the manipulation tool to its end effector and implements manipulation following the trajectory programmed in the last step.

It should be noted that the repeatability accuracy of an industrial robot is assumed to be good in these autonomous teaching applications. However, the proposed framework does have its limitations due to the fact that the on-line accumulated error should be observable by the high-speed sensory feedback in order to realize dynamic compensation.

5.2 Extendibility of the add-on module

Under the condition that the add-on module should have a much higher bandwidth than that of a companion main robot, the add-on module can be designed to be more than two DOFs. Specifically, it is suitable to adopt the parallel mechanism for an add-on module with more than three DOFs to assure motion accuracy. The high-speed sensory feedback for detection of on-line accumulated error can be high-speed vision with 2D/3D configuration or other forms of sensors depending on a specific task.

6. Conclusion

In order to improve the autonomy of industrial robots while at the same time keeping good accuracy under many uncertainties, we present a dynamic compensation framework under a hierarchical intelligent architecture based on a coarse-to-fine strategy. Traditional industrial robots are designated to conduct coarse global motion by focusing on planning-level intelligence. Concurrently, an add-on robotic module with high-speed actuators and high-speed sensory feedback is controlled to realize fine local motion with the role of implementing action-level intelligence. Main advantages of the proposed framework are summarized as follows:

1. Easy calibration for sensor-motor mapping as the industrial robot is designated for coarse motion

2. Good compatibility for industrial robots due to the fact that the control scheme for an arbitrary industrial robot and the add-on compensation module is separated

3. Good compatibility for artificial intelligence algorithms as the planning-level intelligence and action-level intelligence are decoupled.

Author details

Shouren Huang[1]*, Yuji Yamakawa[2] and Masatoshi Ishikawa[1]

1 Graduate School of Information Science and Technology, University of Tokyo, Tokyo, Japan

2 Interfaculty Initiative in Information Studies, University of Tokyo, Tokyo, Japan

*Address all correspondence to: huang_shouren@ipc.i.u-tokyo.ac.jp

References

[1] Verdonck W, Swevers J. Improving the dynamic accuracy of industrial robots by trajectory pre-compensation. In: Proceedings of IEEE International Conference on Robotics and Automation. 2002. pp. 3423-3428

[2] Maeda Y, Nakamura T. View-based teaching/playback for robotic manipulation. ROBOMECH Jounal. 2015;**2**:1-12

[3] Cheah C, Lee K, Kawamura S, Arimoto S. Asymptotic stability of robot control with approximate Jacobian matrix and its application to visual servoing. In: Proceedings of 39th IEEE Conference on Decision and Control. 2000. pp. 3939-3944

[4] Cheah C, Hirano M, Kawamura S, Arimoto S. Approximate Jacobian control for robots with uncertain kinematics and dynamics. IEEE Transactions on Robotics and Automation. 2003;**19**(4):692-702

[5] Zergeroglu E, Dawson D, Queiroz M, Setlur P. Robust visual-servo control of robot manipulators in the presence of uncertainty. Jounal of Field Robotics. 2003;**20**(2):93-106

[6] Piepmeier J, McMurray G, Lipkin H. Uncalibrated dynamic visual servoing. IEEE Transactions on Robotics and Automation. 2004;**20**(1):143-147

[7] Wang H, Liu Y, Zhou D. Adaptive visual servoing using point and line features with an uncalibrated eye-in-hand camera. IEEE Transactions on Robotics. 2008;**24**(4):843-856

[8] Bauchspiess A, Alfaro S, Dobrzanski L. Predictive sensor guided robotic manipulators in automated welding cells. Journal of Materials Processing Technology. 2001;**109**(1–2):13-19

[9] Lange F, Hirzinger G. Predictive visual tracking of lines by industrial robots. International Journal of Robotics Research. 2003;**22**(10–11):889-903

[10] Namiki A, Nakabo Y, Ishii I, Ishikawa M. 1ms sensory-motor fusion system. IEEE Transactions On Mechatoronics. 2000;**5**(3):244-252

[11] Industry 4.0. Available from: https://en.wikipedia.org/wiki/Industry_4.0 [Accessed: 04 May 2016]

[12] Huang S, Yamakawa Y, Senoo T, Ishikawa M. Dynamic compensation by fusing a high-speed actuator and high-speed visual feedback with its application to fast peg-and-hole alignment. Advanced Robotics. 2014;**28**(9):613-624

[13] Huang S, Bergström N, Yamakawa Y, Senoo T, Ishikawa M. Applying high-speed vision sensing to an industrial robot for high-performance position regulation under uncertainties. Sensors. 2016;**16**(8:1195):1-15

[14] Huang S, Yamakawa Y, Senoo T, Bergström N. Dynamic compensation based on high-speed vision and its application to industrial robots. Journal of the Robotics Society of Japan. 2017;**35**(8):591-595

[15] Huang S, Bergström N, Yamakawa Y, Senoo T, Ishikawa M. Robotic contour tracing with high-speed vision and force-torque sensing based on dynamic compensation scheme. In: Proceedings of 20th IFAC World Congress. 2017. pp. 4702-4708

[16] Huang S, Bergström N, Yamakawa Y, Senoo T, Ishikawa M: High-performance robotic contour tracking based on the dynamic compensation concept. In: Proceedings

of IEEE International Conference on Robotics and Automation; 2016. pp. 3886-3893

[17] Sharon A, Hogan N, Hardt E. The macro/micro manipulator: An improved architecture for robot control. Robotics and Computer-Integrated Manufacturing. 1993;**10**(3):209-222

[18] Lew J, Trudnowski D. Vibration control of a micro/macro-manipulator system. IEEE Control Systems Magazine. 1996;**16**(1):26-31

[19] Photron IDP-Express R2000 High-Speed Camera. Available from: http://www.photron.co.jp/ [Accessed: 04 May 2016]

[20] CUDA Parallel Computing Platform. Available from: http://www.nvidia.com/object/cuda_home_new.html [Accessed: 12 June 2016]

[21] Huang S, Yamakawa Y, Senoo T, Ishikawa M. A pre-compensation fuzzy logic algorithm designed for the dynamic compensation robotic system. International Journal of Advanced Robotics. 2015;**12, 12**(3):1

[22] Hartley R, Zisserman A. Multiple View Geometry in Computer Vision. Cambridge, UK: Cambridge University Press; 2004

[23] Lawler E, Lenstra J, Rinnooy K, Shmoys D. The Traveling Salesman Problem: A Guided Tour of Combinatorial Optimization. Chichester, UK: John Wiley and Sons Ltd.; 1985

[24] Video for Peg-and-Hole Alignment Task. Available from: http://www.k2.t.u-tokyo.ac.jp/fusion/pih/index-e.html [Accessed: July 20, 2019]

[25] Huang S, Shinya K, Bergström N, Yamakawa Y, Yamazaki T, Ishikawa M. Dynamic compensation robot with a new high-speed vision system for flexible manufacturing. International Journal of Advanced Manufacturing Technology. 2018;**95**(9–12):4523-4533

[26] Yamazaki T, Katayama H, Uehara S, Nose A, Kobayashi M, Shida S, et al. A 1ms high-speed vision chip with 3D-stacked 140GOPS column-parallel PEs for spatio-temporal image processing. In: Proceedings of International Solid-State Circuits Conference. 2017. pp. 82-83

[27] Video for Contour Tracing Task. Available from: http://www.k2.t.u-tokyo.ac.jp/fusion/tracing/tracing.mp4 [Accessed: 20 July 2019]

Chapter 5

Cooperative Step Climbing Using Connected Wheeled Robots and Evaluation of Remote Operability

Hidetoshi Ikeda, Natsuko Muranaka, Keisuke Sato and Eiji Nakano

Abstract

The present study evaluates the remote operability of step climbing using two connected robots that are teleoperated by individual operators. In general, a teleoperated robot is manipulated by an operator who is viewing moving images from a camera, which is one of the greatest advantages of such a system. However, robot teleoperation is not easy when a teleoperated robot is affected by the force from another robot or object. We constructed a step climbing system using two connected teleoperated robots. A theoretical analysis and the results of simulations clarified the correlations among the robot velocity, the manipulation time of the robots, and the height of the front wheels when climbing a step. The experimental results demonstrate the step climbing ability of the teleoperated robot system.

Keywords: cooperative step climbing, step climbing, wheeled robot, teleoperation, remote operability

1. Introduction

A wheeled mechanism can be easily controlled on a flat road and excels in energy efficiency. Many wheelchairs, carts, and robots that are used in offices or houses have wheeled mechanisms. On the other hand, they face problems with the steps that are commonly found in living spaces. Wheeled mechanisms that can navigate steps or stairs have been widely researched. Such studies have examined additional legs [1, 2], a combination of an adjustable center of gravity and multiple wheels [3], special wheels [4, 5], hopping robots [6], additional driving wheel systems [7], and multiple robots that have forklift mechanisms for climbing steps [8].

We have previously reported cooperative step climbing [9] and descending [10] using two wheeled robots, and we studied step climbing using a wheelchair and a wheeled robot connected by a passive link [11]. We also investigated wheelchair step climbing support by a partner robot equipped with dual manipulators [12, 13]. The above studies were conducted using autonomous or teleoperated robots, both of which have merits and demerits. It is therefore necessary to construct a robot system that is most appropriate for the desired purpose.

The purpose of the present paper is to evaluate the performance of two connected wheeled robots that are teleoperated by individual operators. Teleoperated robots can have cameras and manipulators on their bodies that allow them to be controlled by an operator. They may also be controlled by viewing images obtained from an external camera. This ability to be controlled by humans is one of their advantages.

However, the operators need to pay close attention when robots must be operated with pinpoint precision. For example, when teleoperated robots cooperate with each other to transport an object, interactive forces caused by delays in operation act on the robots. The robots must incline to climb a step, and the interactive forces between robots are always changing. This causes errors in movement or step climbing. Thus, step climbing under control by two operators is not simple. This is therefore the subject of the present study.

The remainder of this paper is organized as follows. Section 2 describes the cooperative system, and Section 3 describes the process of climbing a step. Section 4 presents a theoretical analysis and the results of simulations. Section 5 describes the experiment and results, and Section 6 presents a discussion of the experiments. Finally, Section 7 presents the conclusion of the present paper.

2. Cooperative step climbing robot system

The two robots used in the present study were wheeled robots that were developed in our laboratory (**Figure 1**). Each robot has a pair of front wheels and a pair of rear wheels. The front wheels are casters, and the rear wheels are driving wheels. The robots are connected by a link mechanism, referred to herein as a passive link, and the connecting link positions have free joints. Both robots have mechanisms to change the heights of the connecting positions of the passive link. This step climbing method is affected by the positions of the passive link (see Section 3), and we determined the suitable link positions to overcome a step [14]. The robots are deployed in a forward-and-aft configuration using the link. In the present paper, we refer to the front robot as *Robot A* and the rear robot as *Robot B*. **Figure 2** shows a schematic of cooperative step climbing (see Appendix, **Tables A1** and **A2**).

Figure 3(a) and **(b)** shows the configuration of *Robots A* and *B*. The motors mounted on the robots are connected to a microcontroller (PIC16F873) through a motor driver circuit. The microcontrollers are connected to the robot's PC using a ZigBee module for wireless communications.

The robots have ball screw mechanisms to climb a step and are able to change the link height (see Section 3, **Figure 4**). *Robot A* has touch sensors on its back. The sensors detect the stopping position of the passive link when the operator of *Robot A* controls the link height (**Figure 4**). In the present study, the ball screw mechanism

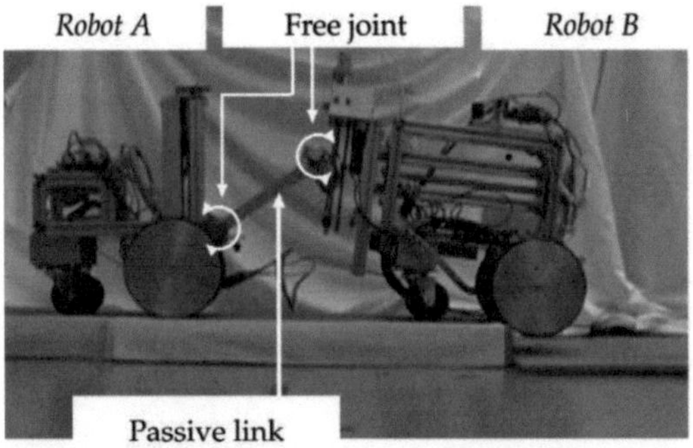

Figure 1.
Robots connected by passive link.

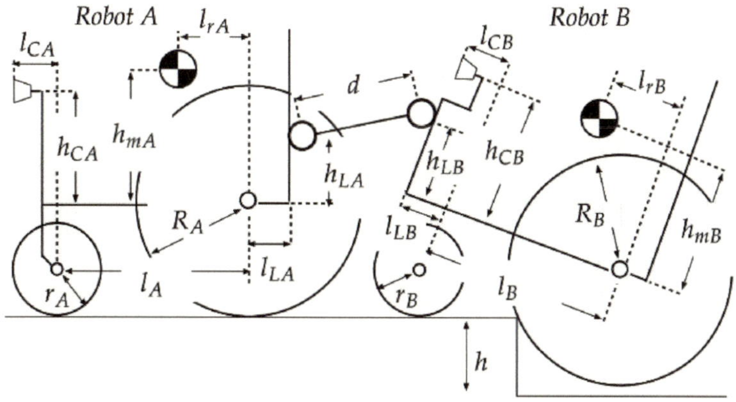

Figure 2.
Schematic of cooperative step climbing and descending robots.

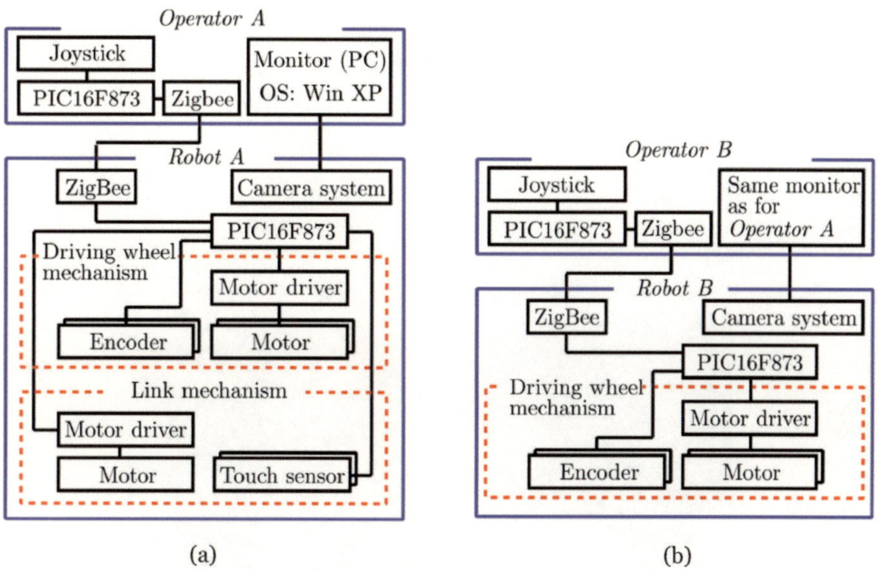

Figure 3.
Configuration of (a) Robot A *and* Operator A *and (b)* Robot B *and* Operator B.

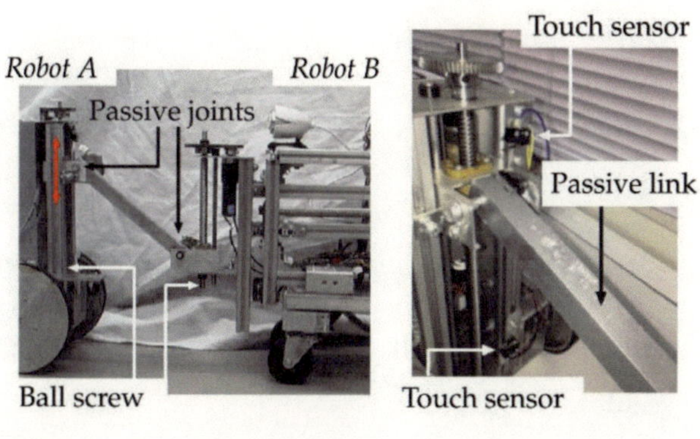

Figure 4.
Passive link mechanism.

of *Robot A* was only used, and the operator of *Robot A* was able to control the link position using a joystick (**Figure 5**).

Each robot was teleoperated by one operator. In the present paper, we refer to the operator of *Robot A* as *Operator A* and the operator of *Robot B* as *Operator B*. The operators controlled each robot using a joystick (**Figure 5**). The robots did not have a system to communicate the information for step climbing; however the operators were able to talk to each other.

Both robots had a camera (ELECOM UCAM-E130HWH, maximum resolution: 1280 × 1024, frame rate: 30 fps (640 × 480 pixels), 10 fps (1280 × 1024 pixels)) on their front (**Figure 6**). The cameras are connected to the PC using USB cables, and

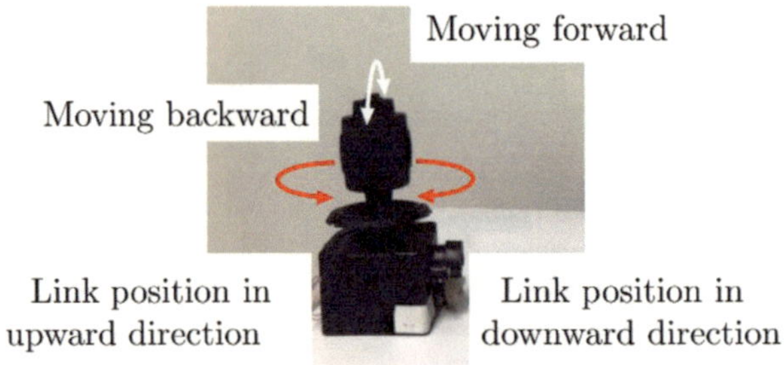

Figure 5.
Joystick for manipulating Robot A.

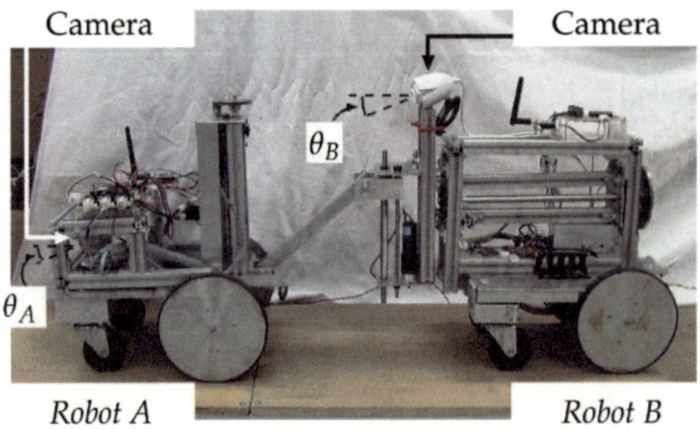

Figure 6.
Cameras on Robot A *and* Robot B.

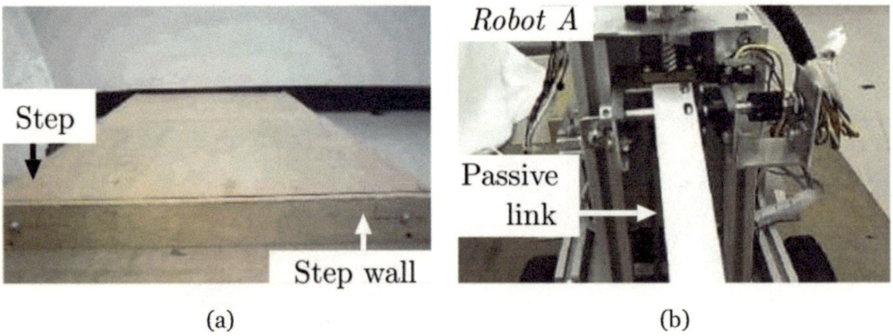

Figure 7.
Moving images from (a) Robot A *and (b)* Robot B.

Figure 8.
PC screen for robot operators.

moving images from both cameras (**Figure 7(a)** and **(b)**) were displayed on the *PC screens* used by the operators (**Figure 8**).

The angles θ_A and θ_B in **Figure 6** are the angles of the cameras on *Robot A* and *B*, respectively. These angles were set to $\theta_A = 10°$ and $\theta_B = 30°$, which are the angles between the robot and the step that the robots are able to see when *Robot A* or *B* inclines to climb a step. *Operators A* and *B* teleoperated each robot using only video data from the cameras.

3. Cooperative step climbing method

The proposed method uses the equilibrium of the robots during step climbing. The two connected robots climb a step sequentially. In the present study, stages 1 and 2 indicate the processes in which the front and rear wheels, respectively, of *Robot A* climb the step. Similarly, stages 3 and 4 signify the processes in which the front and rear wheels, respectively, of *Robot B* climb the step (**Figure 9**). The ascent process, as shown in <1>−<16> in **Figure 9**, is described below. The velocities of the robots are constant in the forward direction.

3.1 Stage 1

<1> Both operators perceive the step using the moving images from the cameras on the robots (**Figure 7**). The link height of *Robot A* is set at a high position (**Figure 4**). <2> *Robot B* stops, and *Robot A* moves forward. As a result, the front wheels of *Robot A* are lifted. <3> The operators make both robots move forward while the front wheels of *Robot A* are lifted. <4> When the operators recognize that the front wheels of *Robot A* have passed over the step edge, the operators manipulate the joysticks to adjust the difference in speed between the robots, so that the front wheels of *Robot A* are placed on the upper level of the step. Here, in stages 1 and 2, if *Robot A* is faster than *Robot B*, then the tilt of *Robot A* increases. If *Robot B* is faster than *Robot A*, then the tilt of *Robot A* decreases.

3.2 Stage 2

<5> The operators make the robots continue to move forward. The back wheels of *Robot A* come into contact with the step. <6> *Robot B* pushes *Robot A*. *Robot B*

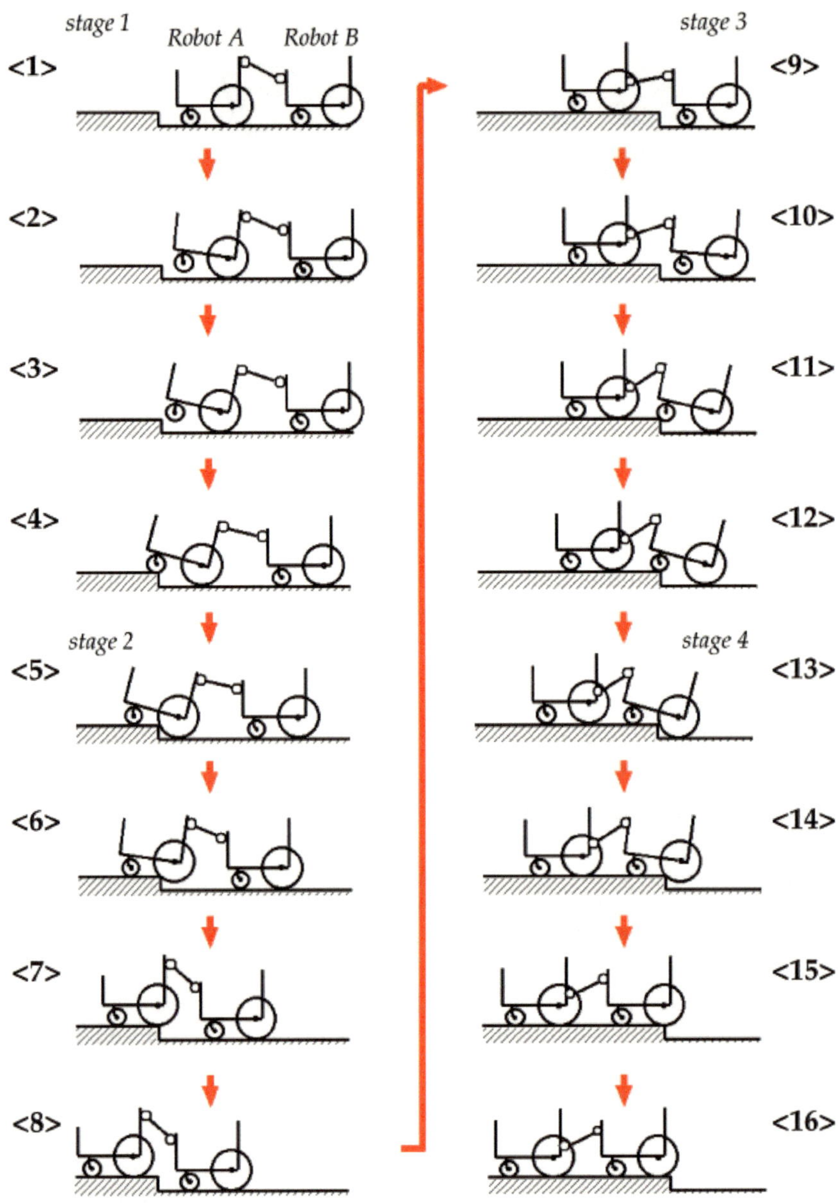

Figure 9.
Entire process of step climbing.

supports the climbing of *Robot A*, and *Robot B* prevents *Robot A* from tipping over backward. <7> *Robot A* climbs up onto the step. <8> Once the rear wheels of *Robot A* have reached the upper level of the step, the operators stop each robot.

3.3 Stage 3

<9> After *stage 2*, the link height of *Robot A* is set at a low position (**Figure 6**). <10> *Operator A* makes *Robot A* stop, and *Operator B* makes *Robot B* move forward. As a result, the front wheels of *Robot B* are lifted. <11> Both robots move forward. <12> The operators recognize that the front wheels of *Robot B* have passed over the step edge. The operators manipulate each joystick to adjust the difference between

the speeds of the robots, so that the front wheels of *Robot B* are placed on the upper level of the step. Here, in stages 3 and 4, if *Robot B* is faster than *Robot A*, the tilt of *Robot B* increases. If *Robot A* is faster than *Robot B*, then the tilt of *Robot B* decreases.

3.4 Stage 4

<13> The operators make the robots continue to move forward. The back wheels of *Robot B* come into contact with the step. <14> *Robot A* pulls *Robot B*. *Robot A* supports the climbing of *Robot B* and *Robot A* prevents *Robot B* from tipping over backward. <15> *Robot B* climbs up onto the step. <16> Once the rear wheels of *Robot B* have reached the upper level of the step, the operators stop each robot.

4. Theoretical analysis

When *Robot A* climbs a step, the body of the robot inclines, and its front wheels are lifted due to the difference in velocity between the two connected robots. When the robots are manipulated by the operators, the step climbing ability greatly influences the manipulation time.

In this section, we clarify the relationships among the robot incline, the velocity, and the manipulation time.

4.1 Relationships among the manipulation time, the velocity, and the height of the front wheels required to climb the step

In **Figure 10**, Σ_B is the basic coordinate system for the robots, where p_0 is the origin as well as the contact position between the rear wheels of *Robot B* and the ground. In addition, $p_i (i = 1–5)$ are the joints (p_1, axis of the rear wheels of *Robot B*; p_2, link position of *Robot B*; p_3, link position of *Robot A*; p_4, axis of the rear wheels of *Robot A*; p_5, axis of the front wheels of *Robot A*; p_6, tread position of the front wheels of *Robot A*).

The position vectors for the joints in the coordinate system Σ_B are expressed as ${}^B\boldsymbol{p}_i = \begin{bmatrix} x_i & y_i \end{bmatrix}^T$ $(i = 1–6)$. In the local coordinate system, for the case in which Σ_i is parallel to Σ_0, ${}^0\boldsymbol{p}_1 = [0 R_B]^T$, ${}^1\boldsymbol{p}_2 = [l_B + l_{LB} h_{LB}]^T$, ${}^2\boldsymbol{p}_3 = [d \quad 0]^T$, ${}^3\boldsymbol{p}_4 = [l_{LA} \quad -h_{LA}]^T$, ${}^4\boldsymbol{p}_5 = [l_A \quad -R_A + r_A]^T$, and ${}^5\boldsymbol{p}_6 = [r_A \quad 0]^T$ (**Figure 10**).

Then, ϕ_i is the angle between Σ_i and Σ_{i-1}, and Σ_1 is parallel to Σ_0 in *stage 1*.

Thus,

$$\phi_1 = 0 \tag{1}$$

The incline of *Robot A*, $\sum_{k=1}^{3} \phi_i$, is

$$\sum_{k=1}^{3} \phi_i = \phi_2 + \phi_3 \tag{2}$$

Here, Σ_4 is always parallel to Σ_3:

$$\phi_4 = 0 \tag{3}$$

In the basic coordinate system Σ_B, the homogeneous transformation matrix ${}^B\boldsymbol{T}_4$ is as follows:

$$
{}^{B}\boldsymbol{T}_4 : \begin{bmatrix} \cos\phi_{23} & -\sin\phi_{23} & x_4 \\ \sin\phi_{23} & \cos\phi_{23} & y_4 \\ 0 & 0 & 1 \end{bmatrix} \tag{4}
$$

Here, $\phi_{23} = \phi_2 + \phi_3$:

$$
x_4 = l_{LA}\cos\phi_{23} + h_{LA}\sin\phi_{23} + l_B + l_{LB} + d\cos\phi_2 \tag{5}
$$

$$
y_4 = l_{LA}\sin\phi_{23} - h_{LA}\cos\phi_{23} + h_{LB} + R_B + d\sin\phi_2 \tag{6}
$$

Then, y_4 is equal to R_A (the radius of the rear wheels of *Robot A*) in *stage 1* (**Figure 10**), and we obtain the following equation from (6):

$$
R_A = l_{LA}\sin\phi_{23} - h_{LA}\cos\phi_{23} + h_{LB} + R_B + d\sin\phi_2 \tag{7}
$$

In this system

$$
R_A = R_B. \tag{8}
$$

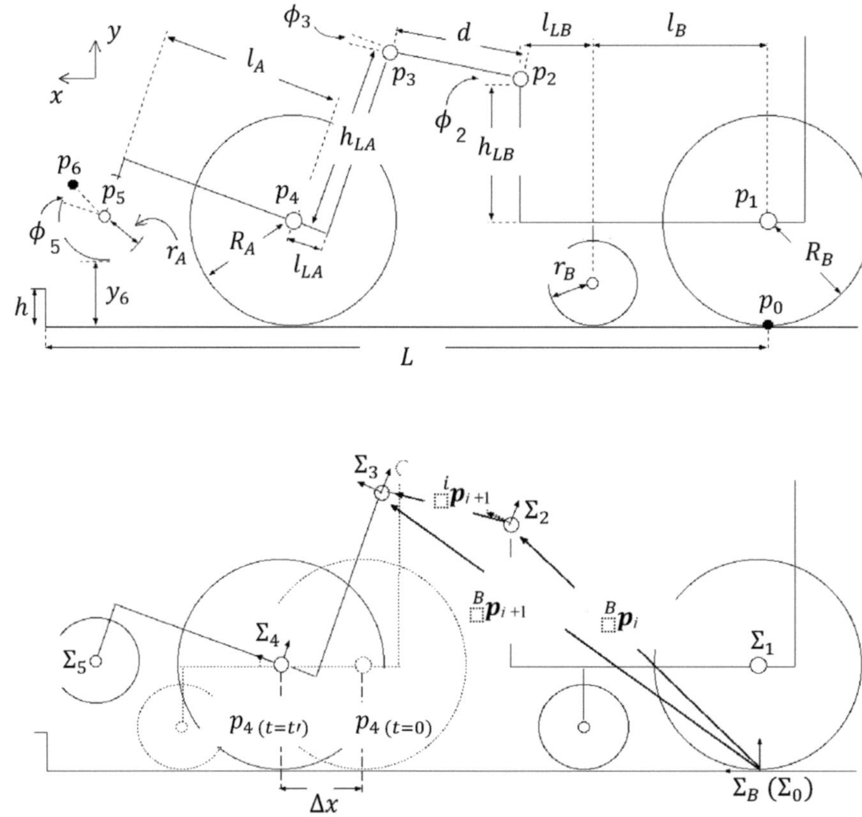

Figure 10.
Lifting the front wheels of Robot A *(stage 1).*

Thus, we have:

$$\sin\phi_2 = \frac{h_{LA}\cos\phi_{23} - l_{LA}\sin\phi_{23} - h_{LB}}{d} \tag{9}$$

Here

$$\cos\phi_2 = \sqrt{1 - \sin^2\phi_2}. \tag{10}$$

The homogeneous transformation matrix, ${}^{B}\boldsymbol{T}_6$, is given by

$${}^{B}\boldsymbol{T}_6 : \begin{bmatrix} \cos\phi_{2356} & -\sin\phi_{2356} & x_6 \\ \sin\phi_{2356} & \cos\phi_{2356} & y_6 \\ 0 & 0 & 1 \end{bmatrix} \tag{11}$$

Here, $\phi_{2356} = \phi_2 + \phi_3 + \phi_5 + \phi_6$.

From (1) and (3), when p_6 (the tread position of the front wheels of *Robot A*, ${}^{B}\boldsymbol{p}_6 = [x_6 y_6]^T$, **Figure 10**) is at the bottom of the front wheels, $\sum_{k=1}^{5}\phi_i = \phi_2 + \phi_3 + \phi_5 = \phi_{235} = -90°$, where

$$\cos\phi_{235} = 0 \tag{12}$$

and

$$\sin\phi_{235} = -1 \tag{13}$$

Thus, we have:

$$\begin{aligned} x_6 &= l_A\cos\phi_{23} - (-R_A + r_A)\sin\phi_{23} + l_{LA}\cos\phi_{23} \\ &\quad + h_{LA}\sin\phi_{23} + l_B + l_{LB} + d\cos\phi_2 \end{aligned} \tag{14}$$

$$\begin{aligned} y_6 &= -r_A + l_A\sin\phi_{23} + (-R_A + r_A)\cos\phi_{23} + l_{LA}\sin\phi_{23} \\ &\quad - h_{LA}\cos\phi_{23} + h_{LB} + R_B + d\sin\phi_2. \end{aligned} \tag{15}$$

By substituting (9) for (15), we have:

$$y_6 = l_A\sin\phi_{23} + (-R_A + r_A)\cos\phi_{23} + R_B - r_A \tag{16}$$

where

$$\sin\phi_{23} = \sqrt{1 - \cos^2\phi_{23}} \tag{17}$$

From (8), (16), and (17), we obtain:

$$\cos\phi_{23} = \frac{e_1\cdot(e_1 - y_6) + l_A\sqrt{l_A{}^2 + e_1{}^2 - (e_1 - y_6)^2}}{l_A{}^2 + e_1{}^2} \tag{18}$$

Here, $e_1 = R_A - r_A$.

Then, p_4 (the position of the axis of the rear wheels of *Robot A*) moves forward in lifting the front wheels of *Robot A* (**Figure 10**). Time $t = tm$ is the time for the operators to lift the wheels.

$p_{4\,(t=0)} = \left[x_{4\,(t=0)} \quad y_{4\,(t=0)}\right]^T$ is the first position of p_4. When the robot manipulation time is $t = 0$, the tilt of *Robot A* is zero ($\phi_{23} = 0$). Then, $p_{4\,(t=0)}$ moves to $p_{4\,(t=tm)} = \left[x_{4\,(t=tm)} \quad y_{4\,(t=tm)}\right]^T$ after *Robot A* moves at a constant velocity of v_A in t s, and Δx is the distance between $p_{4\,(t=0)}$ and $p_{4\,(t=tm)}$.

When the operator begins to teleoperate the robots ($t = 0$), the position of the axis of the rear wheels of *Robot A*, $x_{4\,(t=0)}$, is obtained from (5), as follows:

$$x_{4\,(t=0)} = l_{LA} + l_B + l_{LB} + d\cos\phi_{2\,(t=0)} \tag{19}$$

Here, $\cos\phi_{2\,(t=0)}$ is the value of $\cos\phi_2$ at $t = 0$ s.

After manipulating the robots for $t = tm$, the position of the axis of the rear wheels of *Robot A* ($x_{4\,(t=tm)}$) is given by

$$x_{4\,(t=tm)} = l_{LA}\cos\phi_{23\,(t=tm)} + h_{LA}\sin\phi_{23\,(t=tm)} + l_B + l_{LB} + d\cos\phi_{2\,(t=tm)} \tag{20}$$

Here, $\cos\phi_{23\,(t=tm)}$ is the value of $\cos\phi_{23}$ at $t = tm$.

Based on (19) and (20), the movement distance of *Robot A* while lifting the front wheels, Δx (**Figure 10**), is given as

$$\begin{aligned}\Delta x &= v_A\, t = x_{4\,(t=tm)} - x_{4\,(t=0)}\\ &= l_{LA}\left\{\cos\phi_{23(t=tm)} - 1\right\} + h_{LA}\sin\phi_{23(t=tm)} + d\left\{\cos\phi_{2\,(t=tm)} - \cos\phi_{2(t=0)}\right\}\end{aligned} \tag{21}$$

Here, v_A is the constant velocity of *Robot A*. When the front wheels begin to be lifted ($t = 0$, **Figure 10**), the incline of *Robot A* is zero ($\phi_2 + \phi_3 = 0$). In this case, from (9) and (10), we obtain:

$$\sin\phi_{2\,(t=0)} = \frac{h_{LA} - h_{LB}}{d} \tag{22}$$

and

$$\cos\phi_{2\,(t=0)} = \sqrt{\frac{d^2 - (h_{LA} - h_{LB})^2}{d}} \tag{23}$$

After manipulating the robots for a time tm, the front wheels start to lift. Then, from (9) and (10), we obtain the following:

$$\sin\phi_{2(t=tm)} = \frac{h_{LA}\cos\phi_{23\,(t=tm)} - l_{LA}\sin\phi_{23\,(t=tm)} - h_{LB}}{d} \tag{24}$$

and

$$\cos\phi_{2\,(t=tm)} = \sqrt{\frac{d^2 - {e_2}^2}{d}} \tag{25}$$

Here,

$$e_2 = h_{LA} \cos \phi_{23\ (t=tm)} - l_{LA} \sin \phi_{23\ (t=tm)} - h_{LB}. \tag{26}$$

From (17) and (18), we obtain $\sin \phi_{2(t=tm)}$ and $\cos \phi_{2\ (t=tm)}$, as follows:

$$\sin \phi_{23\ (t=tm)} = \sqrt{1 - \cos^2 \phi_{23\ (t=tm)}} \tag{27}$$

and

$$\cos \phi_{23\ (t=tm)} = \frac{e_1 \cdot (e_1 - y_6) + l_A \sqrt{l_A{}^2 + e_1{}^2 - (e_1 - y_6)^2}}{l_A{}^2 + e_1{}^2} \tag{28}$$

Substituting (22) and (23) for (21), we obtain:

$$t = \frac{1}{v_A}\left[l_{LA}\left(\cos \phi_{23\ (t=tm)} - 1\right) - \sqrt{d^2 - (h_{LA} - h_{LB})^2} + \sqrt{d^2 - e_2{}^2} + h_{LA} \sin \phi_{23\ (t=tm)}\right]. \tag{29}$$

From (26)–(29), the relationships among the velocity of *Robot A* (v_A), the height of the bottom of the front wheels (y_6) (**Figure 11(a)**), and the manipulation time for lifting the front wheels ($t = tm$) are clarified.

4.2 Range of front wheel height within which operators must teleoperate for step climbing

In *stage 1*, the operators lift the bottom of the front wheels of *Robot A* above the step height in order to place the wheels on the step (**Figure 11(a)**). Next, the operators maintain a suitable inclination for *Robot A* in order to prevent it from tipping over backward (**Figure 11(b)**). When *Robot B* stops and *Robot A* continues to move forward, *Robot A* tips over backward (**Figure 12**). Hence, the operators must maintain the height of the front wheels within the range between the step height and the height at which *Robot A* tips over backward.

When the horizontal position of the center of gravity is at the contact position between the rear wheels and the road (**Figure 11(b)**), the height of the bottom of the front wheels is $y_6 = 0.1526$ m. On the other hand, *Robot B* does not tip over backward during *stage 3* because the link touches its body when the incline of *Robot B* grows large (**Figure 13**), so stopping further inclination. The operators are able to teleoperate *Robot B* without tipping the robot over backward.

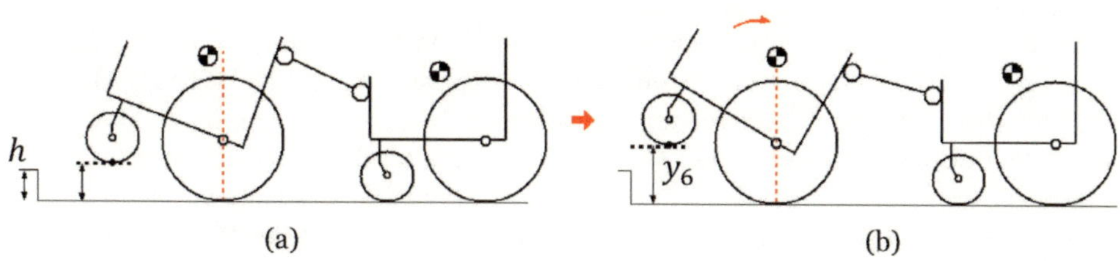

Figure 11.
Height range of the front wheels of Robot A *required to climb a step: (a) situation in which the bottom of the front wheels is equal to the step height and (b) situation in which the horizontal position of the center of gravity is at the contact position between the rear wheels and the road.*

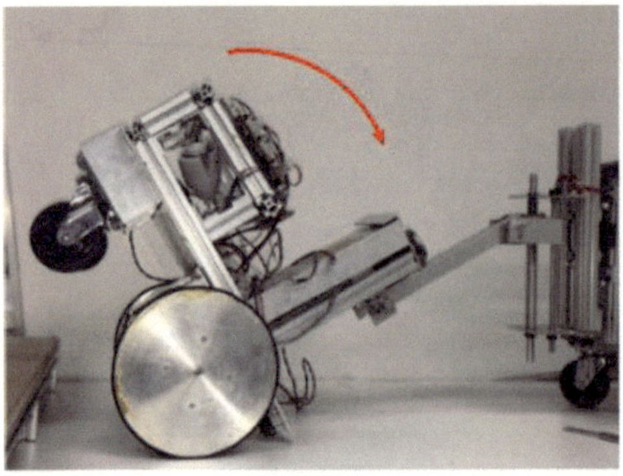

Figure 12.
Tipping over backward of Robot A.

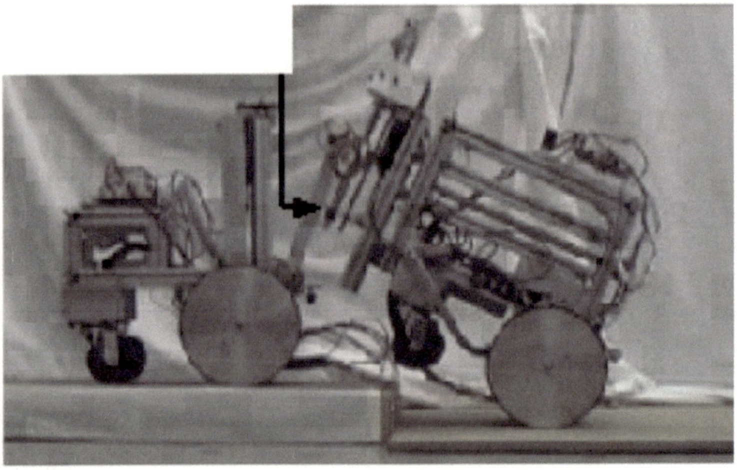

Figure 13.
Link touching the front part of Robot B, *which acts to prevent incline of* Robot A.

4.3 Simulation

We performed a numerical calculation to clarify the correlations among the manipulation time to lift the front wheels, the robot velocity, and the height of the front wheels using (29).

The horizontal axes in **Figure 14(a)**–**(c)** show the manipulation time for lifting of the front wheels, $t = tm$, and the vertical axes show the height of the bottom of the front wheels (y_6), indicating the correlations when the velocity of *Robot A* is $v_A = 0.3$, 0.4, and 0.5 km/h, respectively.

When *Robot A* moves at 0.3 km/h and climbs a 0.05-m-high step, 0.228 s is required to lift the bottom of the front wheels to the step height (**Figure 11(a)**), and 0.559 s is the time when the horizontal position of the center of gravity of *Robot A* is the contact position between the rear wheels and the road surface (**Figure 11(b)**). Therefore, *Operator A* must complete the lifting operation in a time between 0.228 and 0.559 s. If *tm* is less than 0.228 s, the bottom of the front wheels (y_6) will not reach a height of 0.05 m, and if *tm* is greater than 0.559 s, *Robot A* will tip over backward. In **Figure 14(a)**, t_3, which 0.331 s, is the time between these two events. Thus, operators teleoperate the robots to lift the front wheels of *Robot A* and must stop the incline after 0.331 s.

DOI: http://dx.doi.org/10.5772/intechopen.90162

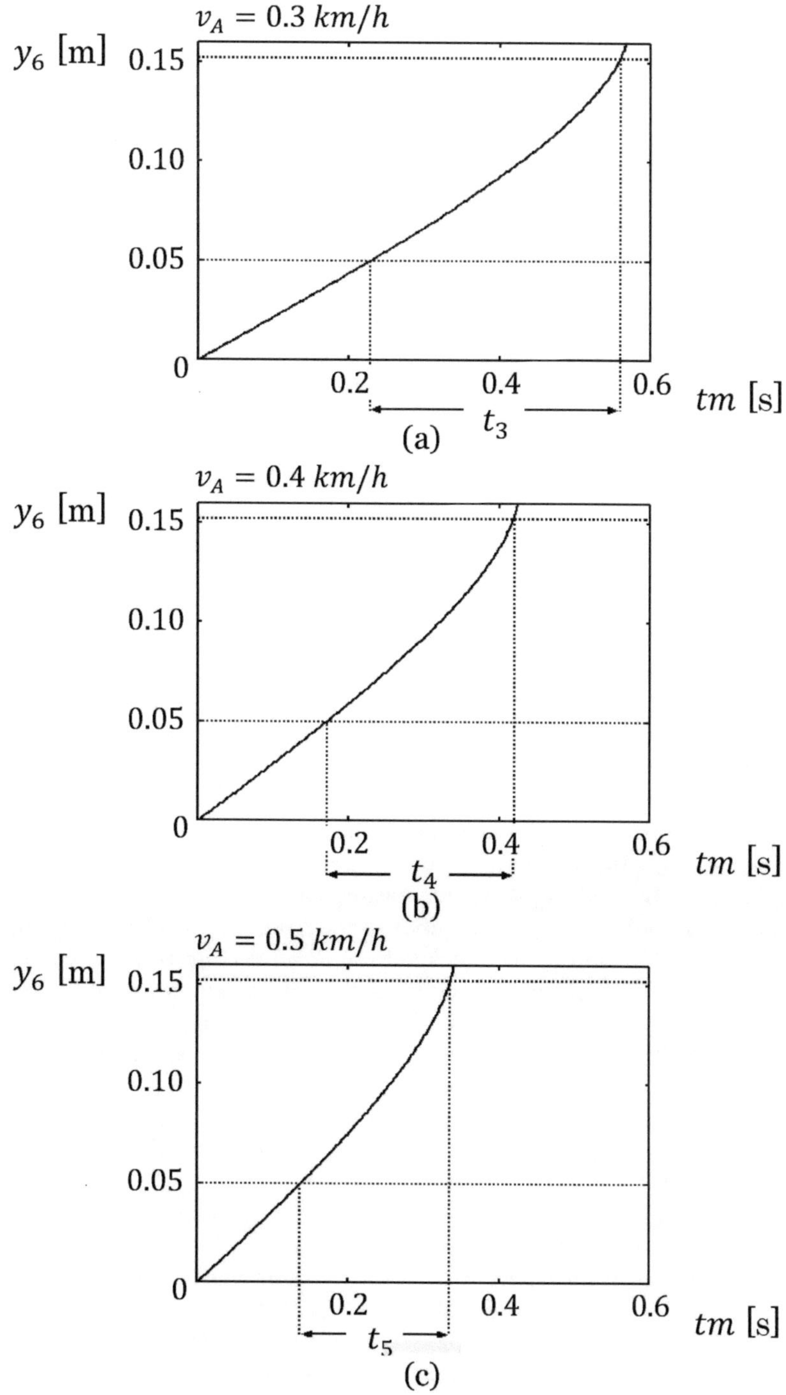

Figure 14.
Relationships among the manipulation time ($t = tm$), the velocity of Robot A (v_A), and the height of the front wheels of Robot A *(y_6) for (a) $v_A = 0.3$ km, (b) $v_A = 0.4$ km, and (c) $v_A = 0.5$ km/h.*

Similarly, t_4 and t_5 are the times at which *Robot A* moves at 0.4 and 0.5 km/h, respectively, where $t_4 = 0.248$ s and $t_4 = 0.199$ s. The results for t_3 and t_5 reveal that t_3 is more than 66.33% of t_5.

In Sections 5 and 6, we discuss the influence of the velocity difference for manipulation of the robots.

5. Experiment

The two robots were teleoperated by individual operators (**Figure 15**). Using joysticks (**Figure 5**), the robots were moved at speeds that were set by the program. Six subjects (five adult males and an adult female) participated as robot operators in the experiments.

The six subjects were labeled s_1 to s_6 and were divided into three groups, α, β, and γ. Subjects s_1 and s_2 were the operators in group α, subjects s_3 and s_4 were the operators in group β, and subjects s_5 and s_6 were the operators in group γ. Subjects s_1, s_3, and s_5 operated *Robot A*, and subjects s_2, s_4, and s_6 operated *Robot B*.

Three experiments were conducted, in which the robot velocities were 0.3, 0.4, and 0.5 km/h. The step height and the friction coefficient were constant at $h = 0.05$ m and $\mu = 0.72$.

The subjects understood the step climbing process and learned how to teleoperate the robots before the experiments. The subjects repeated the test 20 times for each experiment, as was explained before the experiments. When either robot was unable to climb the step, the reason for the failure was recorded. The postures of the robots were then corrected, and the operators restarted the test.

The case in which either of the robots was not able to climb the step was taken to be a step climbing failure. The case in which the both robots were able to climb the step was taken to be a step climbing success.

Tables 1–3 show the results of the experiments for the cases in which the moving robot velocities were 0.3, 0.4, and 0.5 km/h, respectively. The numbers listed in the tables are the test numbers when the robots failed to climb the step, and Col.AS, Tip.A, Col.BS, and Tip.B are the reasons for failure. Here, Col.AS, Tip.A, Col.BS, and Tip.B indicate a collision between the front wheels of *Robot A* and the step wall (**Figure 16**), tipping over backward of *Robot A* (**Figure 12**), a collision between the front wheels of *Robot B* and the step wall, and tipping over backward of *Robot B*, respectively. However, as a result of the link touching its body, tipping over backward of *Robot B* did not occur in the experiments (see **Figure 13**).

In the first experiment (**Table 1**: velocity, 0.3 km/h; success, 52 times; failure, eight times), the success rates for groups α, β, and γ were 85, 85, and 90%, respectively. The reason for failure for all of the groups was a collision between the front wheels of *Robot A* and the step wall.

In the second experiment (**Table 2**: velocity, 0.4 km/h; success, 55 times; failure, five times), the success rates for groups α, β, and γ were 100, 75, and 100%, respectively. The reason for failure for all of the groups was collision between the front wheels and the step wall (*Robot A*, four times; *Robot B*, one time). In the third experiment (**Table 3**: velocity, 0.5 km/h; success, 51 times; failure, nine times), the success rates for groups α, β, and γ were 75, 80, and 100%, respectively.

The reasons for failure for the groups were collision between the front wheels of *Robot A* and the step wall (one time), tipping over of *Robot A* (four times), and collision between the front wheels of *Robot B* and the step wall (four times).

Table 4 lists the ratios for the reasons for failure of the robots to climb the step. The total number of failures for group α was 8 (out of 60 tests), and the total number of failures for groups β and γ were 12 and 2, respectively (**Tables 1–3**).

The most common reason for failure is collision between the front wheels of *Robot A* and the step (59.09%). The second most common reason is collision between the front wheels of *Robot B* and the step (22.73%). Therefore, approximately 82%

DOI: http://dx.doi.org/10.5772/intechopen.90162

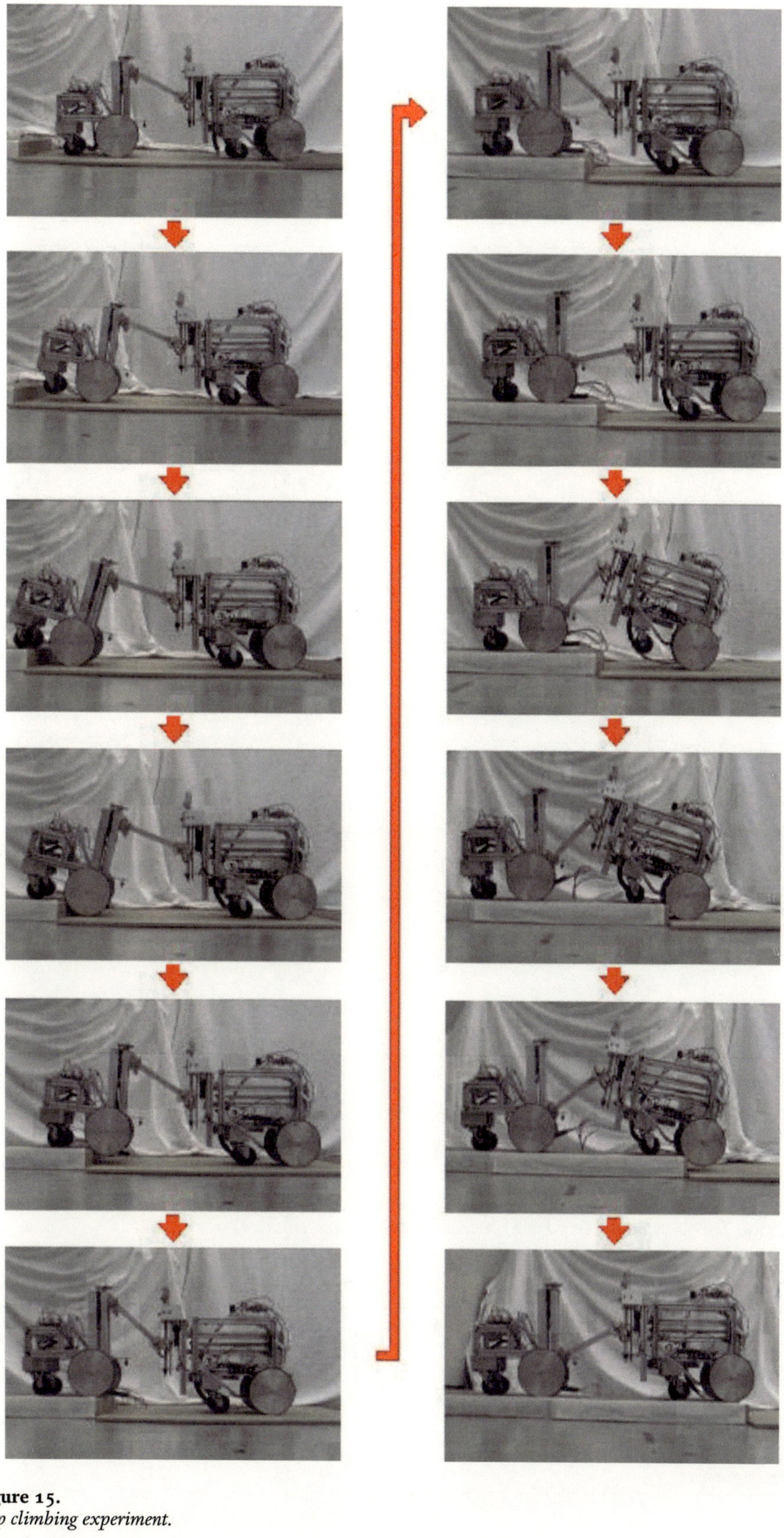

Figure 15.
Step climbing experiment.

Group	Success rate (%)	Reason for failure and test number			
		[Col. AS]	[Tip.A]	[Col.BS]	[Tip.B]
α	85	3	—	—	—
		11	—	—	—
		15	—	—	—
β	85	1	—	—	—
		8	—	—	—
		11	—	—	—
γ	90	3	—	—	—
		4	—	—	—

Table 1.
Success rate for step climbing (0.3 km/h), reason for failure, and test number.

Group	Success rate (%)	Reason for failure and test number			
		[Col. AS]	[Tip.A]	[Col.BS]	[Tip.B]
α	100	—	—	—	—
β	75	2	—	—	—
		3	—	—	—
		7	—	—	—
		—	—	11	—
		20	—	—	—
γ	100	—	—	—	—

Table 2.
Success rate for step climbing (0.4 km/h), reason for failure, and test number.

Group	Success rate (%)	Reason for failure and test number			
		[Col. AS]	[Tip.A]	[Col.BS]	[Tip.B]
α	75	1	—	—	—
		—	2	—	—
		—	—	4	—
		—	—	7	—
		—	—	8	—
β	80	—	—	14	—
		—	18	—	—
		—	19	—	—
		—	20	—	—
γ	100	—	—	—	—

Table 3.
Success rate for step climbing (0.5 km/h), reason for failure, and test number.

Group (total number of failures)	Ratio of reason for failure			
	[Col. AS] (%)	[Tip.A] (%)	[Col.BS] (%)	[Tip.B]
α (8)	50	12.5	37.5	0
β (12)	58.33	25	16.67	0
γ (2)	100	0	0	0
Three groups	59.09	18.18	22.73	0

Table 4.
Ratio of reason for failure of the robots to climb the step (0.3–0.5 km/h).

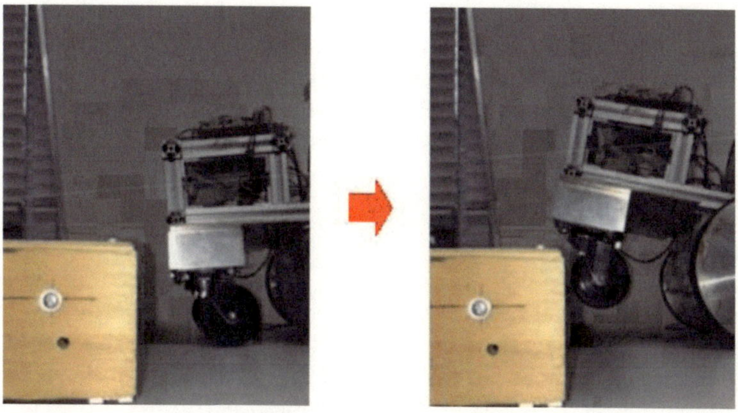

Figure 16.
Collision between the front wheels and the step wall.

of failures arise from collisions between the front wheels and the step. In other words, if a robot is fitted with an assistance system that is able to detect the distance between the robot and the step, the capabilities of teleoperated robots should be greatly improved.

6. Discussion

In this section, we discuss the remote operability of the proposed system based on the results of experiments.

6.1 Correlation between the robot velocity and the success rate

Figure 17(a) shows the correlation between the robot velocity and the success rate for step climbing for the three experiments by group (**Tables 1–3**). The success rate for group γ was high as a whole and increased to 100% at 0.4 and 0.5 km/h. In contrast, the success rate for group α increased at 0.4 km/h but decreased at 0.5 km/h. The success rate for group β decreased at 0.4 km/h and increased at 0.5 km/h. Therefore, the trends in the experimental results are not consistent.

Figure 17(b) shows the total success rate at 0.3–0.5 km/h for the three groups (**Tables 1–3**). The results of the experiments were different from what we had expected, and the remote operability of the step climbing system did not depend on

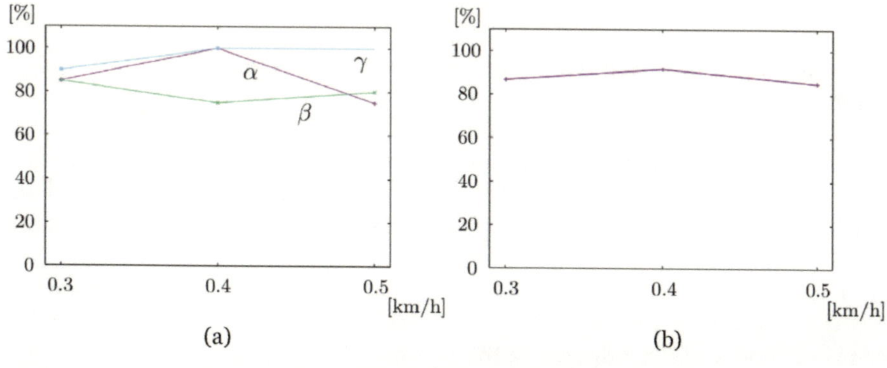

Figure 17.
Ratio of successful step climbing: (a) success rate for each group and (b) total success rate for the three groups.

the velocity of the robots. If the velocity is sufficiently higher than 0.5 km/h, the success rate is likely to be reduced. The operators had to manipulate the robots hurriedly in the experiment in which the velocity was 0.5 km/h. However, the step climbing method was based on the premise that the robots move slowly and in balance with each other and using fast robots is beyond the scope of the proposed method.

6.2 Teleoperation skill

As mentioned above, all subjects knew the step climbing procedure before the experiments and became sufficiently proficient at teleoperating the robots. The experiments were carried out at velocities of 0.3, 0.4, and 0.5 km/h, and 20 tests were performed for each velocity. Thus, each subject performed a total of 60 tests. **Tables 1–3** show that the failures in the experiments by the three groups occurred not only in the early stages of each experiment (1st–7th test) but also in the final stages (14th–20th test). If the subjects did not have sufficient skill to operate the robots, failures should frequently have occurred in the early stages of the experiments at 0.3 km/h. The skill of the operators would be expected to improve as the tests progressed, thus reducing the failure rate in the later tests at each velocity. Also, if the subjects did not have enough skill to react to the speed of the robot, failures should frequently have occurred in the early stages of each experiment (1st–7th test). However, from the results (**Tables 1–3**), failures also occurred in the final stages (14th–20th test), and the success rate was not improved except for group γ (**Figure 17(a)**). Thus, as far as these experiments are concerned, it is clear that the reason for failure was not lack of operator skill.

6.3 Lack of information during teleoperation

Based on interviews with subjects after the experiments, one reason for failure was a lack of information while teleoperating the robots. In these experiments, each robot had one camera on its body, and each subject teleoperated a robot while viewing moving images. The inclines of the robot cameras were set such that the subjects were always able to view the moving images of *Robot A* or the step (**Figure 7(a)** and **(b)**). The subjects had to piece together the status of the climbing robots based on information from the moving images.

We conducted an additional experiment in which a fixed external camera was installed that could view both robots at the same time (see Appendix B, **Figures B1** and **B2**). This experiment was performed using only a single group. However, based on the results, we are fairly certain that using an external camera is effective for teleoperation.

6.4 Losing concentration during teleoperation and conversation between operators

Table 5 lists the total success rate for step climbing for each group (0.3–0.5 km/h). The results for group γ were the best (96.67%), and the results for group β were the worst (80%). There is a difference of approximately 17% for the total success rate between groups γ and β, which is not small.

In the experiment in which the velocity of the robots was 0.5 km/h, the operators had to teleoperate the robots hurriedly, and as a whole did not have enough time to converse while manipulating the robots.

The subjects in group γ (s_5 and s_6) were continuously conversing with each other throughout the experiments and frequently talked about their actions and what

Group	Total success rate for step climbing (%)
α	86.67
β	80
γ	96.67

Table 5.
Total success rate for step climbing for each group (0.3–0.5 km/h).

they intended to do next. As such, conversation was judged to make up for the lack of information during teleoperating the robots.

In contrast, the subjects of group α (s_1 and s_2) conversed little with their partners after the second experiment ($v_A = 0.4$ km/h) because they felt that they were skillful operators and were able to manipulate the robots without conversation. In addition, these subjects were tired from repeating the experiments. The subjects of group β (s_3 and s_4) had few conversations throughout the experiments.

6.5 Results of the experiments

Based on the results of the experiments, it is clear that the cooperative step climbing method can be performed using teleoperated robots. However, it seems reasonable to assume that the ability of the robot system was greatly influenced by loss of concentration and conversation between the robot operators. Even if the operators had sufficient skill to manipulate the robot, they sometimes became tired, and did not converse. The construction of an assist system for manipulation should improve the step climbing ability.

7. Conclusion

The present paper described cooperative step climbing using two wheeled robots connected by a passive link. We constructed a teleoperated robot system and carried out experiments. Our conclusions are as follows:

1. The cooperative step climbing method is practical even if the robots are controlled by teleoperation.

2. A theoretical analysis and the results of simulations clarified the correlations among the manipulation time for the robot, the velocity of the robot, and the height of the front wheels in climbing a step.

3. The ability of operators who reached sufficient proficiency in teleoperating the robots does not depend on the velocity of the robots.

4. It was difficult to perform teleoperation using only moving images from the cameras on the robots because the operators were not able to recognize the overall status of the robots during step climbing.

5. Approximately 82% of the step climbing failures were due to collisions between the front wheels and the step. If the robots have an assistance system that can detect the distance between the robot and a step, the capabilities of such teleoperated robots should improve greatly.

6. Loss of concentration by the operators greatly influenced the operation. Even if the operators had sufficient skill to manipulate the robot, when they became

tired, the success rate for step climbing decreased. The robots are connected by a link and are affected by the force exerted by each other. Therefore, when one or both operators lose concentration, the robots are not able to ascend the step.

7. It is reasonable to assume that conversation between the operators made up for lack of information during teleoperation.

In the future, we intend to construct an augmented reality system to improve remote operability and to perform experiments to confirm its validity. In addition, we will construct an autonomous robot that has sensors and stereo cameras.

Appendix A

Overall length	315 mm
Overall height	395 mm
Radius of front wheels (r_A)	45 mm
Radius of rear wheels (R_A)	80 mm
Wheelbase (l_A)	190 mm
Position of gravity center (l_{rA})	82 mm
Height of gravity center (h_{mA})	74 mm
Link position from rear axle (l_{LA})	65 mm
Height of link position (h_{LA})	40–240 mm
Camera height (h_{CA})	40 mm
Camera position (l_{CA})	95 mm
Mass (M_A)	11.2 kg
Length of link (d)	200 mm

Table A1.
Specifications of Robot A.

Overall length	315 mm
Overall height	365 mm
Radius of front wheels (r_B)	45 mm
Radius of rear wheels (R_B)	80 mm
Wheelbase (l_B)	190 mm
Position of gravity center (l_{rB})	92 mm
Height of gravity center (h_{mB})	75 mm
Link position from front axle (l_{LB})	50 mm
Height of link position (h_{LB})	160 mm
Camera height (h_{CB})	300 mm
Camera position (l_{CB})	75 mm
Mass (M_B)	9 kg

Table A2.
Specifications of Robot B.

Appendix B

Group α conducted 20 tests (maximum velocity 0.3 km/h). **Table B1** lists the experimental results. It can be seen that the success rate was improved from 85 (**Table 1**) to 95%. Based on interviews with subjects s_1 and s_2 (operators in group α), teleoperation in this manner was easier than using cameras on the robots. Although the experimental results were obtained using only one group, we are fairly certain that using an external camera is effective for teleoperation. If the environment does not allow for the placement of an external camera, the robots should have a sensor system that can show the status of the robots and the step for assistance in teleoperation.

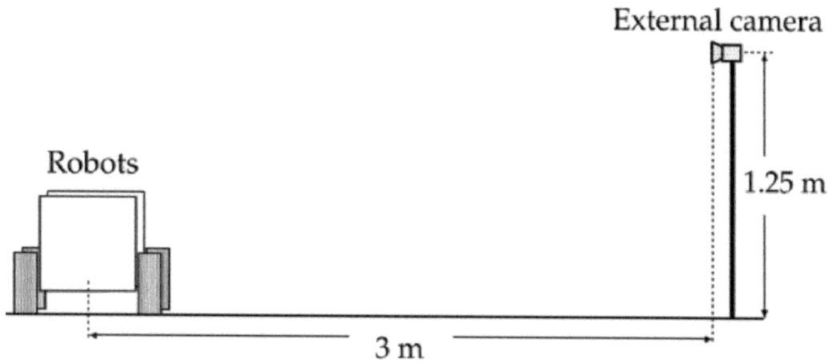

Figure B1.
A schematic of the setup used in an experiment in which an external camera (iBuffalo BSW3KMW01, maximum frame rate: 30 fps) was used to view both the robots and the step together.

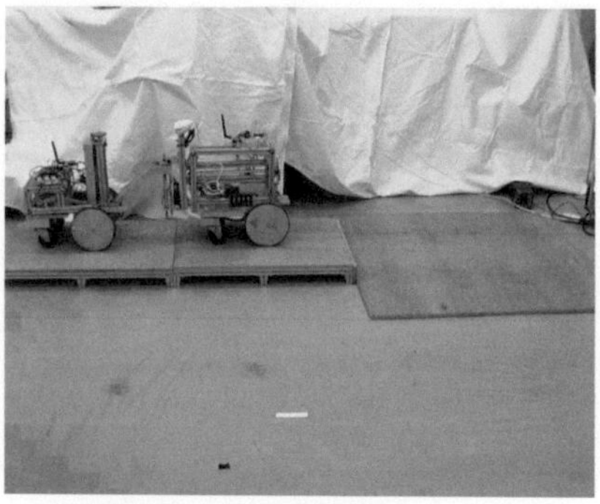

Figure B2.
The robots did not have a camera on their body in this case, but the operators were able to determine the status of the robots using the external camera.

Group	Success rate	Reason for failure and test number			
		[Col. AS]	[Tip.A]	[Col.BS]	[Tip.B]
α	95%	18th	—	—	—

Table B1.
Ratio of reason for failure of the robots to climb the step (0.3–0.5 km/h).

Author details

Hideotoshi Ikeda[1*], Natsuko Muranaka[1], Keisuke Sato[1] and Eiji Nakano[2]

1 National Institute of Technology, Toyama College, Toyama, Japan

2 Robofesta org., Tsukuba, Japan

*Address all correspondence to: ikedah@nc-toyama.ac.jp

References

[1] Kumar V, Krovi V. Optimal traction control in a wheelchair with legs and wheels. In: Proceedings of the 4th National Applied Mechanisms and Robotics Conference, December 1995; Cincinnati, 95-030-01-95-030-07

[2] Nakajima S, Nakano E, Takahashi T. Free gait algorithm with two returning legs of a leg-wheel robot. Journal of Robotics and Mechatronics. 2008;**20**(4): 662-668

[3] Independence Technology, iBOT [Internet]. Available from: https://www.verywellhealth.com/stair-climbing-wheelchair-ibot-mobility-system-187962

[4] Taguchi K. Enhanced wheel system for step climbing. Advanced Robotics. 1995;**9**(2):137-147

[5] Eich M, Grimminger F, Kirchner F. A versatile stair-climbing robot for search and rescue applications. In: 2008 IEEE International Workshop on Safety, Security and Rescue Robotics; Sendai, Japan; 2008. p. 35-40

[6] Stoeter S, Rybski P, Gini M, Papanikolopoulos N. Autonomous stair-hopping with scout robots. In: IEEE/RSJ International Conference on Intelligent Robots and Systems; Lausanne, Switzerland; 2002. p. 721-726

[7] Munakata Y, Wada M. Modeling and analysis of static wheelie of a five-wheeled wheelchair for climbing over a step. In: 2014 IEEE/ASME International Conference on Advanced Intelligent Mechatronics; Besanc, France; 2014. p. 8-11

[8] Asama H, Sato M, Goto N, Kaetsu H, Matsumoto A, Endo I. Mutual transportation of cooperative mobile robots using forklift mechanisms. In: IEEE International Conference on Robotics and Automation. Minneapolis, MN, USA; 1996. p. 1754-1759

[9] Ikeda H, Wang Z, Takahashi T, Nakano E. Stable Step Climbing and Descending for Tandem Wheelchairs Connected by a Passive Link. In: IEEE/ICME International Conference on Complex Medical Engineering. Beijing, China; 2007. p. 23-27

[10] Ikeda H, Nagai S, Doba H, Nakano E. Cooperative step descending control which is based on the information from the angle of the link with connecting wheelchair robots (in Japanese). Journal of the Society of Life Support Technology. 2014;**26**(2):64-71

[11] Ikeda H, Katsumata Y, Shoji M, Takahashi T, Nakano E. Cooperative strategy for a wheelchair and a robot to climb and descend a step. Advanced Robotics. 2008;**22**:1439-1460

[12] Ikeda H, Hashimoto K, Murayama D, Yamazaki R, Nakano E. Robot teleoperation support system for collision avoidance between wheelchair front wheels and a step. In: The 2016 IEEE International Conference on Simulation, Modeling, and Programming for Autonomous Robots. San Francisco, CA, USA; 2016. p. 203-209

[13] Ikeda H. Chapter 10: Step climbing strategy for a wheelchair. In: Advances in Intelligent Systems: Review Book Series. Vol. 1. Barcelona, Spain: IFSA Publishing, S.L.; 2017. pp. 249-288

[14] Ikeda H, Wang Z, Takahashi T, Nakano E. Step climbing and descending method by tandem wheelchairs and analysis of the influence of changes of the operator's posture (in Japanese). Journal of the Society of Biomechanisms. 2003;**27**(3): 134-143

Chapter 6

Real-Time Robot Software Platform for Industrial Application

Sanghoon Ji, Donguk Yu, Hoseok Jung and Hong Seong Park

Abstract

In this study, we present the requirements of a real-time robot software (SW) platform that can be used for industrial robots and examine whether various kinds of existing middleware satisfy them. Moreover, we propose a real-time robot SW platform that extends RTMIA to various industrial applications, which is implemented on Xenomai real-time operating system and Linux. The proposed SW platform utilizes the timer-interrupt based approach to keep strict period and the shared memory for convenient usage, on which the shared variable is designed and used. We verify the proposed platform by showing that the robot task and the Programmable Logic Controller (PLC) program are performing with interlocking each other on the presented platform.

Keywords: software platform, real-time, middleware, industrial robot, PLC

1. Introduction

A robot software (SW) platform is an execution environment that provides various functions to easily execute various robot softwares. Therefore, a middleware that provides such functions can be considered as a platform.

Middleware is a type of software that exists between an application and an operating system, and it employs communication protocols on the hardware, thereby facilitating the development applications by users and reducing the cost and risk when developing them [1]. Several types of middleware have been applied in the field of robotics, such as Common Object Request Broker Architecture (CORBA) [2, 3], Real-Time CORBA (RT-CORBA) [2, 4], Data Distribution Service (DDS) [2, 5], OPC Unified Architecture (OPC-UA) [6], Robot Operating System (ROS) [7, 8], Open Platform for Robotic Services (OPRoS) [9–11], OpenRTM (open robotics technology middleware) [12], open robot control software (OROCOS) [13], XbotCore [14], and real-time middleware for industrial automation devices [15] (hereafter, referred to as RTMIA).

A common feature of these middleware is that they support message-based communication. Thus, a majority of middleware can be implemented in multiple nodes, such as processes, threads, or CPU boards that support middleware-unique communication protocol to facilitate data exchange between them. The difference among these middleware is whether the real-time constraint is satisfied or not, from the perspective of the application thread or process. Even in the real-time viewpoint, providing real-time from a communication point of view is different from providing real-time from a (application) process point of view. In general, the term "real-time" in this study indicates the real-time used in the process viewpoint. Hence, middleware must be able to control user-created programs using processes or threads.

Even though middleware such as CORBA, RT-CORBA, OPRoS, openRTM, OROCOS, XbotCore, and RTMIA control threads, in other middleware such as OPC-UA, ROS, and DDS, the control of threads or processes is implemented by the operating system. Hence, they do not directly control threads or processes. Except CORBA, all of the above-mentioned thread controlling middleware satisfy the real time constraint. However, if they are analyzed in detail, RT-CORBA, OPRoS, openRTM, and OROCOS have conditions to satisfy the real time, in which the execution type is limited to thread type and threads can be controlled in the time range such as over 1 ms [8] or 10 ms [11]. In [14] EtherCAT-based robot real-time SW platform is proposed, which is a Xenomai-based robot control middleware that satisfies the hard real-time requirements and is designed to perform 1 kHz control loops in a multi-axis system consisting of 33 axes. RTMIA is middleware that can execute control loops in 100 μs and can simultaneously control both processes and threads in real-time.

OPC-UA, ROS, and DDS are a kind of communication middleware. That is, they do not satisfy the real time requirement since they are in charge of the data exchange part between multiple nodes. Note that ROS 2.0 utilizes DDS. OPC-UA is a middleware that performs data exchange among various robots and controls devices such as the FMS controller or the server on the upper level. OPC-UA is not a middleware that can be used for robot control like ROS. The use cases of ROS shows that most types of the user program are process types rather than thread types. In other words, the thread type provided by the real-time middleware is reluctant to be used because their basic usage forms may introduce inconvenience and unfamiliarity for robot software developers. ROS has big advantage of using the conventional programming method, which users are familiar with, and utilizes the ready-made existing robot program. However, ROS must use 'sleep()' function for periodic processing. 'sleep()' function does not always wake up at the correct time, resulting in time shift occurrence. Accumulation of such shifts cause loss of some periods, which may lead to states that cannot be controlled properly. In order to avoid such situation, some constraints must be kept to strictly control the number of applications and the execution environment.

The current trend of commercial PLC-related products [16, 17] shows that platforms such as CODESYS [16] and TwinCAT [17] control robot systems, including grippers, by embedding robot motion control SWs to PLC SWs. The feature of these platform is that they use the robot block language proposed by PLCopen [18], while their principal advantage is the ability to link robots to related automation devices efficiently using PLC program. However, the types of robots that can be controlled through these platforms are still limited. To solve this, the platform should simultaneously perform both PLC function blocks, including robot function blocks, and robot (task) programs defined directly by users or robot developers.

In this study, we presented the requirements of the real-time robot SW platform for industrial robots and examined whether various types of existing middleware satisfy them. Additionally, we extend the RTMIA [15] and propose a real-time robot SW platform that can be used for various industrial applications and satisfies the presented requirements. We implemented our proposed platform on the Xenomai real-time operating system and Linux, and verify it via a few practical examples, wherein the interlocking programs of robot task and PLC are simultaneously performed.

The remainder of the study is organized as follows. In Section 2, we present the requirements of the real-time robot SW platform for industrial applications and discuss whether various kinds of existing middleware satisfy them. Section 3 presents the implementation of the real-time robot SW platform, and Section 4 describes the practical results obtained by mutually interlocked work of programs of robot task and PLC on the proposed platform. Finally, Section 5 concludes the study.

2. Operation requirements and analysis of real-time robot SW platform

For a real-time robotic SW platform for various industrial applications, we must know the characteristics (period, etc.) of the data used for processing periods and events. Referring to the data presented in [19], the data used in the manufacturing robot can be roughly described as in **Table 1**. It may differ slightly from other classifications because it is categorized based on type and period (or delay time) of control/sensor data.

In fact, the data generated at each level in **Table 1** is used as basis for the target data. For example, upon recognizing an obstacle at level 4, its position data is used to generate a level 3 position profile that provides relevant position data of level 2, based on which, at level 1, a position control loop moves the motor to the desired position by controlling the speed or torque of the motor. That is, data of each level is interlocked with each other in terms of control.

A real-time robot SW platform should satisfy the following requirements.

R1: Support data exchange
R2: Real time support (strict period execution and sporadic performance support)
R3: Supports thread and process types for user defined programs
R4: Easy configuration of applications (robot control SW, PLC SW, vision inspection SW, non-real-time SW, etc.)
R5: Support multiple periods.
R6: Threads or processes running in the same period are classified by priority.
R7: Check and handle the event through the event handler.

R1 serves as a communication middleware by providing a method for exchanging data between configured applications. R2 supports real time operation and requires a minimal amount of jitter to perform periodic operations. It must also be able to handle real-time events. For instance, conditions for use in various control loop programs are strict periodic execution conditions, including the program for directly controlling the motor without using a separate controller, and sporadic to handle emergency event such as an emergency stop. In many cases, such execution condition is required. In particular, the 100 μs period can be used to control the motor current. R3 uses threads for real-time support, but also allows the process to

Level	Data type	Period (delay time)
4	Obstacle recognition, object recognition, device status	Seconds
3	Position profiles	Hundreds of ms
2	Position data	Tens of ms
1	Motor torque and speed control data	100 μs

Table 1.
Types of control/sensor data used in manufacturing robots and their associated periods (delays).

be used for soft real-time execution. Process support, as mentioned earlier, is about using legacy programs. R4 does not control the developed application software modules in a hard-coded program, but rather configures them by freely setting the period, priority, application name, etc. through a configuration file.

R5 and R6 are about the execution period. As shown in **Table 1**, various execution periods are required. For example, in the case of motor control, if the order of sensor reading (RS), control value generation (GC), and motor value writing (WM) are different, the value sampled in the previous period is used. In other words, when RS, GC, and WM are executed, the control value can be read within one period and sent to the motor. However, when operating in the order of GC, RS, and WM, the control value calculated using the value read in the previous period affects the motor. That is, the previous value without using the current motor related data may affect the motor control. Therefore, to avoid such effects, it is necessary to set the execution priority of SW module within the same period so that they can operate in the order of RS, GC, and WM.

R7 relates to how to handle events. Depending on the event behavior utilized by the operating system, there exist interrupt-based and program-based event handling methods. Interrupt-based events can be used if the HW interrupt can be controlled directly, however, modern operating systems prevent such handling for system stability. Therefore, it is common to introduce an event handler to first check whether an event has occurred, and further handle it. Even those manufacturing robots can generate and handle relevant events such as when there is an emergency stop switch, or the maximum value of sensor input or output exceeds a certain limit. Meanwhile, the processing time for these events is also important. Emergency stop switches should be used as quickly as possible because they are related to safety.

The comparison of the results of whether the existing middleware satisfy each of the presented requirements, is presented in **Table 2**.

Requirement number		**ROS**	**RT-CORBA**	**OPRoS**	**openRTM**	**OROCOS**	**XbotCore**	**CODESYS**	**RTMIA**
R1		O	O	O	O	O	O	O	O
R2	General	△[1]	O[2]	O	O	O	O	O	O
	period of 100 μs	X	△ (25 ms)	△ (10 ms)	△ (10 ms)	△ (10 ms)	△ (1 ms)	n.a	O
	R3	△p[3]	O	△t[4]	△t[4]	△t[4]	△t[4]	n.a	O
	R4	△[5]	O	O	O	O	△	O	O
	R5	△[1]	O	O	O	O	O	O	O
	R6	X	X	X	X	X	X	n.a	X
	R7	X	△	X	X	X	X	n.a	X

n.a, not available.
O, support; X, not support; , support under some constraints.
[1]Leverage OS features rather than middleware.
[2]Realtime only supports threads.
[3]p means only process type.
[4]t means only thread type.
[5]All source code is public, but all applications are compiled/linked.

Table 2.
Comparing existing middleware to platform requirements.

3. Implementation and motion analysis of real-time robot SW platform

In general, multiple periods are controlled by using the greatest common divisor (GCD) and the least common multiple (LCM) of the periods. If the period, which is the greatest common divisor, is less than the minimum period provided by the middleware, the multiple periods must be adjusted so that the GCD period is equal to or greater than the minimum period. The smaller the provided period, the better it is because it can be used for various applications. Let the GCD period be the basic period and the LCM period be the macro period [15]. **Table 3** illustrates a periodic scheduling table using GCD/LCM based on **Figure 1**.

The middleware must be implemented to execute the periodic threads and processes, the sporadic threads and processes, as well as the non-real-time processes, as shown in **Figure 1**. Periodic threads and processes should be executed within given periods, while sporadic threads and processes should be executed within the deadline once the corresponding events occur. An example of this behavior is shown in **Figure 2**. Of course, non-real-time modules are executed only when execution time remains in a period. As shown in **Figure 2**, periodic execution has the highest priority, followed by sporadic execution. After the execution of periodic modules, sporadic execution checks whether the event occurs and if true, the event handler invokes the corresponding event processing function to handle it. These operations demonstrate that requirements R2, R3, R5, R6, and R7 are satisfied. Especially, we use time-based interrupts to keep the period strict and calculate the n-th jitter, as shown in Eq. (1) [15]:

$$J_n = P_n - T_n = T_0 + n * period - T_n \tag{1}$$

where

T_0: the first execution time of the target module (reference time)

P_n: start time of the n-th period, indicated by $T_0 + n * period$

T_n: start time of the n-th module

$period$: basic period

From Eq. (1) and **Figure 2**, we can observe that it is important to keep the period (n = 0,1,2, ...) correctly. There are two approaches to implement the periodic execution: use timer interrupt and use the sleep function. The second method is the most frequently used one and its usage can be described as follows: after executing the SW modules, the operating system sleeps during the remaining time until the next period. However, this method has a disadvantage that the remaining time period does not keep up correctly, leading to increasing jitter. Therefore, in this study, we use a method of implementing period based on timer interrupts. That is, the minimum period is the minimum interval of the timer interrupt generated by the operating system. This allows middleware to be designed to meet requirements

No. of basic period	Execution time (ms)	Execution module (in priority)	Macro period
0	0	cntr3 > cntr2 > cntr1 > cntr4	Repeat basic periods every 0.5 ms
1	0.1	cntr3 > cntr2	
2	0.2	cntr3 > cntr2	
3	0.3	cntr3 > cntr2	
4	0.4	cntr3 > cntr2	

Table 3.
*Periodic scheduling table for **Figure 1**.*

```xml
<?xml version="1.0" encoding="UTF-8"?>
<!-- filename: file to be loaded and executed by middleware -->
<!-- moduletype: process(exe type), thread(so or dll type) -->
<!-- operationtype: periodic, sporadic, non-real -->
<!-- period: nano sec -->
<!-- priority: the lowest value is the highest priority -->
<!-- property: input parameters needed to execute the module -->
<root>
  <module> <!-- periodic module with period of 100 us -->
    <filename>./cntr1.so</filename>
    <moduletype>thread</moduletype>
    <operationtype>periodic</operationtype>
    <period>500000</period>
    <priority>3</priority>
    <property>
      <value name="intialize_value">5</value>
    </property>
  </module>
  <module>
    <filename>./cntr2.so</filename>
    <moduletype>thread</moduletype>
    <operationtype>periodic</operationtype>
    <period>100000</period>
    <priority>2</priority>
  </module>
  <module>
    <filename>./cntr3.so</filename>
    <moduletype>thread</moduletype>
    <operationtype>periodic</operationtype>
    <period>100000</period>
    <priority>1</priority>
  </module>
  <module>
    <filename>./cntr4.exe</filename>
    <moduletype>process</moduletype>
    <operationtype>periodic</operationtype>
    <period>500000</period>
    <priority>4</priority>
  </module>
  <module> <!-- sporadic module with deadline of 100 us -->
    <filename>./event1.so</filename>
    <moduletype>thread</moduletype>
    <operationtype>sporadic</operationtype>
    <deadline>1000000</deadline>
    <priority>1</priority>
  </module>
  <module>
    <filename>./visionEvt.exe</filename>
    <moduletype>process</moduletype>
    <operationtype>sporadic</operationtype>
    <deadline>100000000</deadline>
    <priority>2</priority>
  </module>
  <module> <!-- non-real-time module -->
    <filename>./nonReaTime.exe</filename>
    <moduletype>process</moduletype>
    <operationtype>non-real</operationtype>
  </module>
</root>
```

Figure 1.
Example of module.xml.

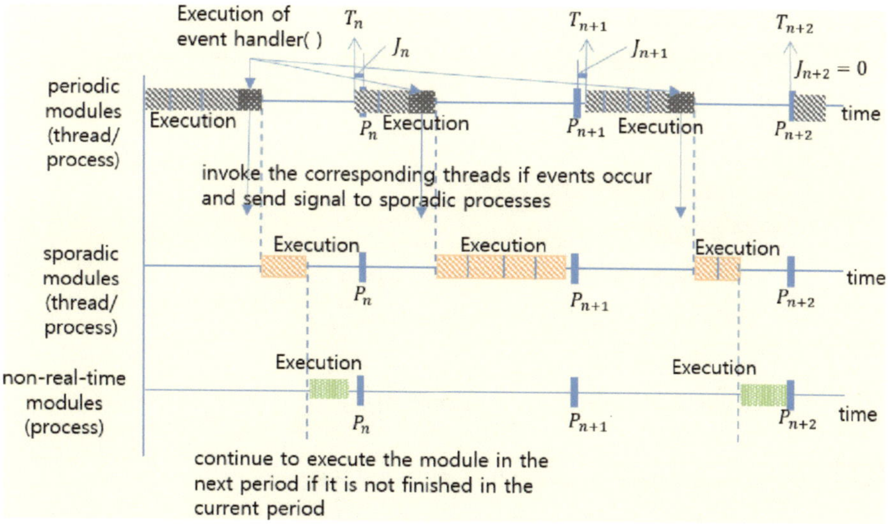

Figure 2.
Example of execution timing diagram of SW modules.

R2 and R5. In addition, the platform should be designed using the POSIX (Portable Operating System Interface) standard.

The real-time robot SW platform should be able to exchange data or remotely perform necessary functions among periodic SW module, sporadic SW module and non-real-time SW module. This is because only cooperation between SW modules can achieve the desired result. For this purpose, the proposed middleware implements a method of obtaining the desired results by performing exchange of data and remote functions using shared memory. The advantage of using shared memory is that it is easy to maintain data consistency and to use the data. In other words, when SW modules including input module write input data to shared memory, other SW modules read them and calculate the control values at the same time. And if necessary, the module writes the result to shared memory for other modules to use the data. In addition, it provides a shared function in the shared memory method so that other SW modules can use the shared functions of the modules. Of course, when using shared memory, the problem of mutual exclusion between threads is solved.

A file named .glb is used to define shared variables that are implemented using shared memory. For example, if a file called knuRobot.glb is created as shown in **Figure 3**, a conversion function creates the header file and a .cpp function to access shared variables. The difference from existing shared variable is that, just like global variables, the shared variable that resides in shared memory can be used in all the software modules. This method makes the program very simple. Example programs that use shared variables in **Figure 3** are shown in **Figures 4** and **5**, in which,

```
// knuRobot.glb
namespace knu.Robot;
uint32_t tick;
struct {
   double x; double y; double z;
} joints[10]
```

Figure 3.
Example code of a file defined shared variables.

```
// output.cpp
#include <indurop/indurop.h> // header file for Middleware
#include <indurop/generated.h> // generated by .glb

using namespace knu::Robot;
void onRun() {
   INDUROP_TYPEOF(joints)::value_type v = joints[0];
   irp::printf("joints[0] = {%lf, %lf, %lf}₩n", v.x, v.y, v.z);
}
INDUROP_MODULE(run = &::onRun)
```

Figure 4.
*Example program 1 that reads and prints shared variable defined in **Figure 3**.*

```
// modify.cpp
#include <indurop/indurop.h> // header file for Middleware
#include <indurop/generated.h> // generated by .glb
using namespace knu::Robot;
void onRun() {
   tick = tick + 10;
   INDUROP_TYPEOF(joints)::value_type v = joints[0];
   v.x += tick; v.y += tick; v.z += tick;
   joints[0] = v;
}
INDUROP_MODULE(run = &::onRun)
```

Figure 5.
*Example program 2 to update shared variables defined in **Figure 3**.*

"#include <indurop/generated.h>" is a header file that should be added when shared variables are used, which is generated from .glb file during preprocessing. Therefore, simply by adding "indurop/generated.h", the variables 'tick' and 'joints [10]' defined in the knuRobot.glb file can be accessed. The INDUROP_MODULE statement in **Figures 4** and **5** represents information about the module's name, description, author, license, and callback function. The middleware periodically executes the module, where the function to be called is specified as "run = &:: onRun." Furthermore, 'run' is the tag name of the periodically called callback function and "&::onRun" defines the address of the function "void onRun()" presented in each figure. Using INDUROP_TYPEOF(argument) in **Figure 5**, the same data type as the variable defined in .glb file can be created and used.

The event handler in **Figure 5** is invoked using the basic period in **Figure 2**. In other words, it is invoked after executing periodic modules and before processing the sporadic service. Event handlers can be implemented using sporadic threads or sporadic processes. The algorithm of the event handler is shown in **Figure 6**.

Figures 7 and **8** show examples of sporadic thread and process-related programs that allow event handling, respectively.

Figure 6 depicts an algorithm that describes the event handler's operation. It can execute the thread program of **Figure 7** and the process program of **Figure 8**. The event handler first calls the thread's 'condition()' function and if the execution result is true, calls the 'run()' function to handle the event. Then it sends signal to the processes to handle the event occurred. In case of thread, event handling function can be specified separately, but, in this study, 'run()' function is used for simplicity, whose example is shown in **Figure 7**. **Figure 8** shows an example of a process that handles an event. In this example, it utilizes signal because the process cannot be called directly from an event handler. **Figures** 7 and **8** also illustrate the use of shared variables.

```
eventHandler()
{
  for (all threads enrolled to event scheduler) {
    call the functions 'condition()' of sporadic threads,
    if (the result of condition() is true)
       invoke run() of the corresponding thread
  }
  for (all processes enrolled to event scheduler) {
    send signal to each process
  }
}
```

Figure 6.
Algorithm for event handler.

```
#include <indurop/generated.h> // generated by .glb
extern "C" int condition(){
    if(joint[0].x >= 180){        // event is occurred
        return 1;
    } else{
        return 0;
    }
}
extern "C" void run(void){
    // process the corresponding event
}
```

Figure 7.
Example code of a sporadic thread program handled by an event handler.

```
//sporadic-process1
#include <indurop/indurop.h>
#include <fstream>

int condition(event){
 if(event <0 || event >= 180)
           return 1;  // occur event
 else
           return 0;
}
int main(int argc, char* argv[]){
    double y_pos;
    channel::Handle handle;   // connection channel for event handler
    irp::initPeriodExe();     // enrollment of signal to event handler


    // connect channel to process for shared variables ;
    ...
   while(){
        irp::waitPeriod();  // wait signal from event handler
        read_ret = read(handle, "joint[1].y", sizeof(y_pos), &y_pos);
          // read value from shared variable named "joint[1].y" in Fig. 3
        if (condition(y_pos))
            processEvent(y_pos) ;
    }
    return 0;
}
void processEvent(double event)
{
  // process event
}
```

Figure 8.
Example code of a sporadic process program handled by an event handler.

4. Robot system integrating robot controller and PLC

The XML file in **Figure 9** is a configuration file consisting of programs of robot task, PLC, and an input/output data processing. As shown in **Figure 9**, the program consists of three periodic threads, where the period of all SW modules is 10 ms. main.so consists of a ladder program, which is converted to C/C++ program, and a C++ program that calls the program, which is shown in **Figure 12**. The robot task program in **Figure 13** is stored in Robot.so. The reason why we only used thread type is that, comparing to process type, it runs more precisely with real-time. The execution order is as follows: the input/output data processing module, the robot task module, and the PLC module. Of course, users can modify the execution order by changing priorities of modules. Once this configuration file is processed, the robot system will operate as shown in **Figure 10**, which can be configured and operated with the process.

Defined in **Figure 11**, shared variables are used in the integrated application system of **Figure 10**, in the PLC ladder program in **Figure 12** and also in the robot control program in **Figure 13**. The name space 'PLC_ROBOT' is used in **Figure 13**, which makes the program look more sophisticated. As shown in **Figures 11** and **13**,

```
<?xml version="1.0" encoding="UTF-8" ?>
<root>
<module>
 <filename>./build/main.so</filename>
 <moduletype>thread</moduletype>
 <operationtype>periodic</operationtype>
 <period>10000000</period>
 <priority>3</priority>
 <property>
   <value name="counter">5</value>
 </property>
</module>
<module>
 <filename>./build/Robot.so</filename>
 <moduletype>thread</moduletype>
 <operationtype>periodic</operationtype>
 <period>10000000</period>
 <priority>2</priority>
 <property>
   <value name="counter">5</value>
 </property>
</module>
<!--module>
<filename>/usr/local/InduRoP/lib/libiodispat.so</filename>
<moduletype>thread</moduletype>
 <operationtype>periodic</operationtype>
 <period>10000000</period>
 <priority>1</priority>
 <property>
   <value name="config">/home/rcmain1/cmd_Demo/ioconfig.xml</value>
 </property>
</module!-->
</root>
```

Figure 9.
Configuration xml file for integrated application system.

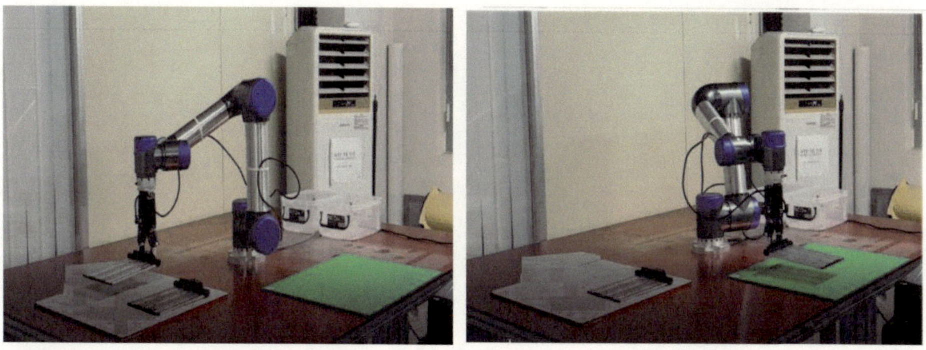

Figure 10.
Results from integrated application systems.

the name space 'PLC_ROBOT' indicates that shared variables are utilized. The PLC ladder program shown in **Figure 12** is converted to C/C++ function named 'POU_PROGRAM()' and made into a thread to run. Through this, various types of

```
namespace PLC_ROBOT;
short I0 // PLC input point 0
short I1 // PLC input point 1
short O0 // PLC output point 0

...... // Code omitted

#define MAX_JOINTS          6
struct {
unsigned short nAXESGROUP;
short MotionDone;
short Busy;
short eState;
short      bRuntimeService;
short      bJob_RobotMoving;            // Motion command interpreting
unsigned long      dCurJoint[6];        // servo current joint angle //// 6 --> MAX_JOINTS
unsigned long      dCmdJoint[6];        // servo command joint angle //// 6 --> MAX_JOINTS
unsigned long      dWorldPos[6];        // world position (command position)
unsigned long      dUserPos[6];         // User position (command position)
unsigned long      dBasePos[6];         // Base position (command position)
unsigned long      dJointSpeed[6]; // 6 --> MAX_JOINTS
unsigned long      dBaseSpeed[6];
unsigned long      dWorldSpeed[6];
unsigned char      nposeINFO[6]; // 6 --> MAX_JOINTS
} ROBOT_MODULE_STATE; // External Information
..
struct
{
int32_t                eOpCode;
unsigned short      t_pl;                          // position level
          unsigned long       t_speed;             // speed
          unsigned short      p_Target;
unsigned short      p_Mid;
} ROBOT_MODULE_SERVICE; // To ROBOT MODULE

...... // Code omitted

struct {
          uint16_t error;
          uint16_t status;
          int8_t mode;
          int32_t position;
} AXIS1_IN;
```

Figure 11.
Shared variables used in integrated application systems.

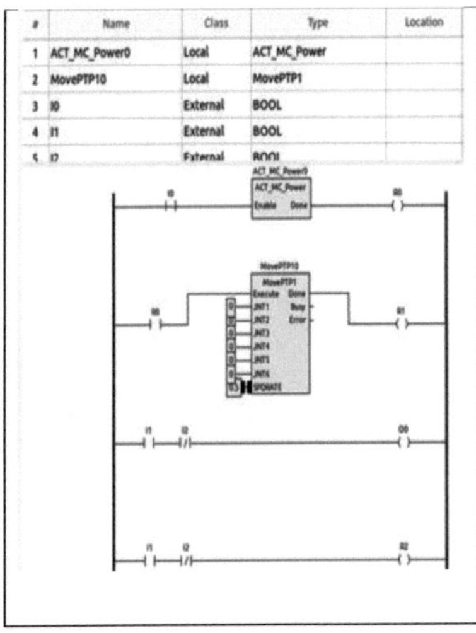


```
... // Code omitted

#include <indurop/indurop.h>
#include <indurop/generated.h>
#include "POUS.h"

extern "C" void initialize(irp::Property
const& property)
{
std::cout <<"[PLC_IO_TEST]" <<std::endl;
}

void start()
{
POU_Initialize();
}

void run()
{
POU_PROGRAM(); //call ladder program
}
.... // Code omitted
```

Figure 12.
PLC ladder program example using shared variable.

```
.. // Code omitted

#include <indurop/indurop.h>
#include <indurop/generated.h>

#include "platform.h"
#include "KinematicFunction.h"
#include "MotionControl.h"
#include "MotionPlan.h"
..
void start()
{
int i, j, k;
int nRetCode;

std::cout <<"[START]" <<std::endl;

PLC_ROBOT::ROBOT_MODULE_STATE.eState = plcRobot.eState = ROBOT_GROUP_DISABLED;
PLC_ROBOT::ROBOT_MODULE_SERVICE.eOpCode = SVC_AD_GROUP_ENABLE;
PLC_ROBOT::msgRobotServiceEditNum=1;
plcRobot.plcRobotInfo.nRobotAxis = 6;

for (k=0; k<6; k++)
PLC_ROBOT::ROBOT_MODULE_STATE.dCurJoint[k] = 0.0;
}

void run()
{
int i, j, k;
int nRetCode;
int32_t joint[6];

joint[0] = PLC_ROBOT::AXIS1_IN.position / 1320000 * 360;
joint[1] = PLC_ROBOT::AXIS2_IN.position / 1320000 * 360;
joint[2] = PLC_ROBOT::AXIS3_IN.position / 1320000 * 360;
joint[3] = PLC_ROBOT::AXIS4_IN.position / 1320000 * 360;
joint[4] = PLC_ROBOT::AXIS5_IN.position / 1320000 * 360;
joint[5] = PLC_ROBOT::AXIS6_IN.position / 1320000 * 360;

for (k=0; k<6; k++)
plcRobot.dCurJoint[k] =PLC_ROBOT::ROBOT_MODULE_STATE.dCurJoint[k] = (unsigned long)

if (PLC_ROBOT::msgRobotServiceEditNum!=0) {
msgRobotService[0].eOpCode=
(OPCODE_ROBOT_SERVICE)((int) PLC_ROBOT::ROBOT_MODULE_SERVICE.eOpCode);
msgRobotService[0].arg.eOpCode = OPCODE_MOVJ;
msgRobotService[0].arg.ARG_CMD.argMOVE.t_speed.dValue= 50;
//:ROBOT_MODULE_SERVICE.t_speed;
nRetCode = Handling_Robot_Service(&msgRobotService[0]);
if(PLC_ROBOT::ROBOT_MODULE_STATE.MotionDone==1)
PLC_ROBOT::msgRobotServiceEditNum = 0;

... // Code omitted
N=5 ; // N is Gear Ratio
PLC_ROBOT::AXIS1_OUT.position = N*1320000 * joint[0] / 360;
PLC_ROBOT::AXIS2_OUT.position = N*1320000 * joint[1] / 360;
PLC_ROBOT::AXIS3_OUT.position = N*1320000 * joint[2] / 360;
PLC_ROBOT::AXIS4_OUT.position = N*1320000 * joint[3] / 360;
PLC_ROBOT::AXIS5_OUT.position = N*1320000 * joint[4] / 360;
PLC_ROBOT::AXIS6_OUT.position = N*1320000 * joint[5] / 360;
}
```

Figure 13.
Robot joint control SW module.

PLC programs can be operated. Threads that link shared variables to I/O modules are defined in the libiodispat.so, which is illustrated in the third part of **Figure 11**, and the middleware executes the thread. If the thread libiodispat.so periodically reads sensing data from the input module and stores into shared variables, then control threads of robot of Robot.so and PLC of main.so access shared variables.

5. Conclusions

In this study, we presented seven requirements of real-time robot SW platform that can be used for industrial robot and examined whether existing middleware such as ROS, OPROS, openRTM, and RTMIA [15] satisfy these requirements. In particular, communication middleware such as ROS has a disadvantage of demanding the user to have more knowledge about the real-time operating system to use the industrial robot but advantage that its usage is simpler because it does not manage execution of processes and/or threads. On the other hand, OPRoS, openRTM, and OROCOS manage the execution of threads for periodic execution but does not control the execution of processes.

In this study, we proposed the real-time robot SW platform that satisfied the presented seven requirements R1–R7 by extending RTMIA and demonstrated its implementation on the Xenomai real-time operating system and Linux. The proposed SW platform utilized the timer-interrupt based approach to keep strict period and the shared memory for convenient usage.

We applied our method to a practical robot system, wherein the programs of PLC and the robot were used simultaneously, and their corresponding operating results were also presented. In this implementation, PLC ladder program and robot control program were managed using period of 10 milliseconds. As the implemented application was simple, it was not shown that event handling and execution of periodic processes were working well. But they did work well. That is, it can be known that the proposed platform satisfied requirements R1–R7. As a result, the platform proposed in this study was verified.

The other advantage of this study is the method to access shared variable. It can be known that it is generally easy and convenient to read and write the shared variables using the conventional variable access method when the shared variables defined in other threads are accessed. Hence the proposed method for accessing of shared variables is designed for multiple processes or threads to have mutually exclusive access to shared variables using the conventional variable access method. And it is shown that the proposed method for accessing of shared variables is working well.

Future work may include investigating a robotic platform that optimally operates multiple threads in multicore systems. And the proposed platform will be implemented on the μC/OS.

Acknowledgements

This work was partially supported by KITECH research fund and Korea Evaluation Institute of Industrial Technology (KEIT) grant funded by the Ministry of Trade, Industry & Energy (KOREA) (No. 20004535 and No. 20005055).

DOI: http://dx.doi.org/10.5772/intechopen.91014

Author details

Sanghoon Ji[1], Donguk Yu[2], Hoseok Jung[1] and Hong Seong Park[3]*

1 Convergent Technology R&D Division, Robot R&D Group, Korea Institute of Industrial Technology, Gyeonggi-do, South Korea

2 Navcours, Daejeon, South Korea

3 Department of Electrical and Electronic Engineering, Kangwon National University, Gangwon-do, South Korea

*Address all correspondence to: hspark@kangwon.ac.kr

References

[1] Mahmoud QH. Middleware for Communication. Chichester: John Wiley and Sons, Ltd.; 2004. DOI: 10.1002/0470862084

[2] Pérez H, Gutiérrez JJ. Mint: A survey on standards for real-time distribution middleware. ACM Computing Surveys. 2014;**46**(4):49. DOI: 10.1145/2532636

[3] OMG. CORBA Core Specification. v3.3 [Internet]. 2012. Available from: https://www.omg.org/spec/CORBA/3.3/

[4] OMG. Real-time CORBA Specification. v1.2 [Internet]. 2005. Available from: https://www.omg.org/spec/RT/About-RT/

[5] OMG. Data Distribution Service for Real-Time Systems. v1.4 [Internet]. 2015. Available from: https://www.omg.org/spec/DDS/

[6] OPC Foundation. OPC Unified Architecture Specification Part 1: Overview and Concepts. 2017. Available from: https://opcfoundation.org/developer-tools/specifications-unified-architecture/part-1-overview-and-concepts

[7] Open Source Robotics Foundation. ROS (Robot Operating System) [Internet]. Available from: www.ros.org

[8] Kay J, Tsouroukdissian AR. "Real-time Performance in ROS 2," ROSCon2015 [Internet]. 2015. Available from: https://roscon.ros.org/2015/presentations/RealtimeROS2.pdf

[9] OPRoS (Open Platform for Robotic Services). [Internet]. Available from: https://github.com/opros-wiki/OPRoS_v1.1_Components/wiki/Open-Platform-for-Robotic-Services

[10] Han S, Kim M, PHS. Mint: Open software platform for robotic services. IEEE Transactions on Automation Science and Engineering. 2012;**9**(3):467-481. DOI: 10.1007/978-3-319-11900-7

[11] Lee D, AH. Mint: Real-time characteristics analysis and improvement for OPRoS component scheduler on windows NT operating system. Journal of Institute of Control, Robotics and Systems (in Korean). 2011;**17**(1):38-46. DOI: 10.5302/J.ICROS.2011.17.1.38

[12] openRTM (open Robotics Technology Middleware). [Internet]. Available from: http://openrtm.org/

[13] OROCOS (Open Robot Control Software) project. [Internet]. Available from: http://www.orocos.org/

[14] Muratore L, Laurenzi A, Hoffman EM, Rocchi A, Caldwell DG, Tsagarakis NG. Mint: On the design and evaluation of XBotCore, a cross-robot real-time software framework. Journal of Software Engineering for Robotics. 2017;**8**(1):164-170. DOI: 10.6092/JOSER_2017_08_01_p164

[15] Park HS. Real-time scheduling method for middleware of industrial automation devices. In: Open Access Peer-Reviewed Chapter—Online First. 2019. DOI: 10.5772/intechopen.86769

[16] CODESYS. [Internet]. Available from: https://www.codesys.com

[17] TWINCAT. [Internet]. Available from: https://www.beckhoff.com/twincat/

[18] PLCopen. [Internet]. Available from: https://plcopen.org/

[19] Meier L, Honegger D, Pollefeys M. Mint: PX4: A node-based multithreaded open source robotics framework for deeply embedded platforms. In: 2015 IEEE International Conference on Robotics and Automation (ICRA); Seattle, WA, USA; 26–30 May 2015. DOI: 10.1109/ICRA.2015.7140074

Chapter 7

Visual-Tactile Fusion for Robotic Stable Grasping

Bin Fang, Chao Yang, Fuchun Sun and Huaping Liu

Abstract

The stable grasp is the basis of robotic manipulation. It requires balance of the contact forces and the operated object. The status of the grasp determined by vision is direct according to the object's shape or texture, but quite challenging. The tactile sensor can provide the effective way. In this work, we propose the visual-tactile fusion framework for predicting the grasp. Meanwhile, the object intrinsic property is also used. More than 2550 grasping trials using a novel robot hand with multiple tactile sensors are collected. And visual-tactile intrinsic deep neural network (DNN) is evaluated to prove the performance. The experimental results show the superiority of the proposed method.

Keywords: stable, grasp, tactile, visual, deep neural network

1. Introduction

In recent years, dexterous robotic manipulation increasingly attracts worldwide attention, because it plays an important role in robotic service. Furthermore, the stable grasp is the basis of manipulation. However, stable grasp is still challenging, since it depends on various factors, such as the actuator, sensor, movement, object, environment, etc. With the development of the neural network, the data-driven methods [1] become popular. For example, Levine et al. used 14 robots to randomly grasp over 800,000 times for collecting the data and training the convolutional neural network (CNN) [2]. Guo et al. trained the deep neural network (DNN) with 12 K-labeled images to learn the end-to-end grasping polices [3]. Mahler et al. built the dataset that included millions of point cloud data for training Grasp Quality Convolutional Neural Network (GQ-CNN) with an analytic metric. Then GQ-CNN developed the optimal grasp strategy that achieves 93% success rate for eight kinds of objects [4–6]. Zhang et al. trained robots to manipulate objects by videos that were made by virtual reality (VR). For pick-place tasks, the success rate was increased when the number of samples increased [7]. Therefore, sufficient high-quality data is important for robotic grasping.

Nowadays, a few datasets of robot grasping have been developed. Playpen dataset obtains 60-hour grasping data of robot PR2 with RGBD cameras [8]. Columbia dataset collects about 22,000 grasping samples via the GraspIt! simulator [9]. Besides experiments with robots and numerical simulations, human manipulation videos are also useful. Self-supervised learning algorithms are developed from demonstration of videos [10]. While the above datasets focus on the whole grasping process, there are other datasets that concentrate on specific tasks, like grasp planning and slip detection. Pinto et al. instructed robots to automatically generate labeled images for grasp planning with 50,000 times by self-supervised learning algorithms [11]. MIT built the grasp dataset by vision-based tactile sensor and external vision [12]. While some experiments produced slip with extra force or fix objects [13, 14], researchers recorded the actual random grasping process with 46% failure results in 1000 times grasp [15, 16]. The real data can contribute to the precision grasping [17]. In daily life overabundance of the object's types leads to the difficulty of building datasets. Some researchers select the common objects and build 3D object set models such as KIT objects [18], YCB object set [19], etc. They are more convenient for research. However, there are few datasets that include the visual and tactile data. Sufficient visual, tactile, and position data can clearly describe the grasping process and improve the robot's ability of grasping.

According to the previous work, it is necessary to build a complete dataset for the robotic manipulation. In this chapter, a new grasp dataset based on the three-finger robot hand is built. In the following section, the structure of the multimodal dataset is introduced in detail. Moreover, the CNN and long short-term memory networks (LSTMs) are designed to complete grasp stability prediction.

2. Grasp stability prediction

In this section, the multimodal fusion framework of grasp stability prediction is proposed.

2.1 Visual representation learning

Under the visual image set, we can only observe 2700*2 = 5400 sets of image data in total, which is in use. It is difficult to extract visual features with convolutional neural networks (ResNet-18 network structure is used in our experiment). Training convergence is less on a small dataset, so time comparison network is used [10], capture video information from the capture process, anchor, positive, negative data. Then we define the triplet loss function [20] and use the characteristics of the continuous change of motion in the video to learn the operation process. The visual characteristics are also used as a pre-training process for the subsequent stable retrieval of the convolutional network part of the prediction network. Such as shown in **Figure 1**, we cleverly use a multi-angle camera to record the video image of the same capture process; at the same time, different image in the perspective should represent the same robot state, that is, its embedded layer embedding vector. A certain distance from the feature representation is relatively small, and the image at the same perspective at different times represents the robot. At different grasping states, a certain distance of the embedded layer Embedding vector is relatively large, formally:

$$\|f(x_i^a) - f(x_i^p)\|_2^2 + \alpha < \|f(x_i^a) - f(x_i^n)\|_2^2 \tag{1}$$

where $f(x_i^a)$, $f(x_i^p)$, and $f(x_i^n)$ represent the anchor, positive, and negative image features extracted by CNN. So, we can define the loss function [21] as

$$l(a,p,n) = \frac{1}{N}\left(\sum_{i=1}^{N} \max\{d(a_i,p_i) - d(a_i,n_i) + \alpha, 0\}\right) \tag{2}$$

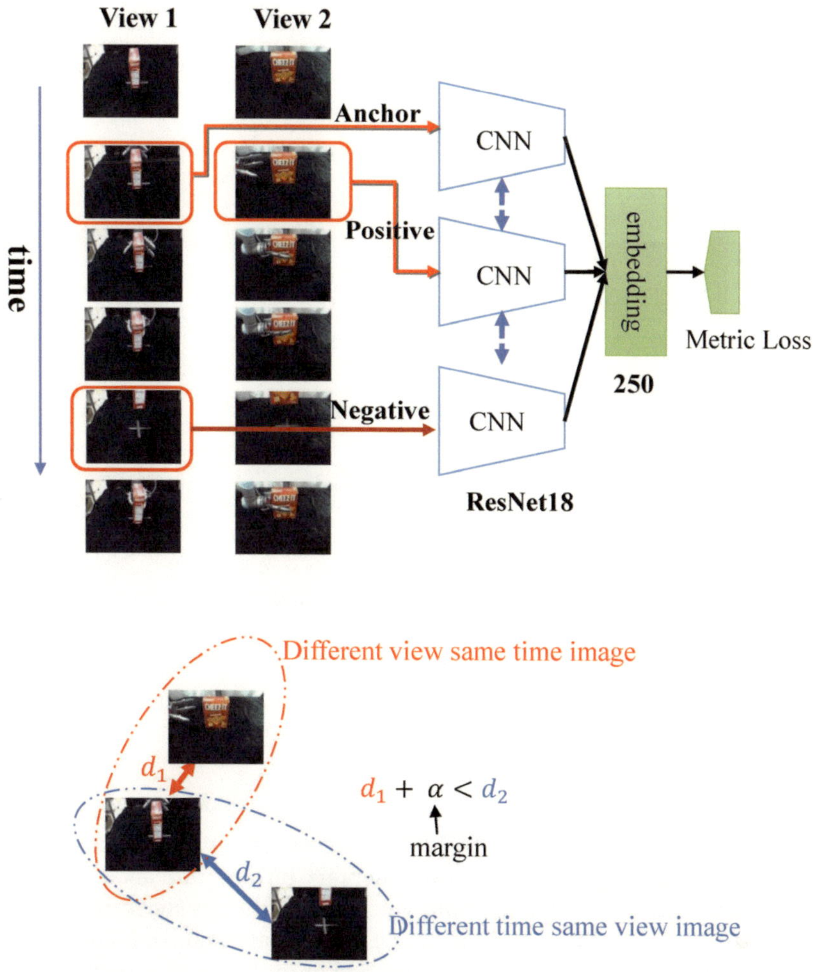

Figure 1.
Visual representation network.

2.2 Predicting grasp stability

In order to describe the properties of the objects like shape or size, the images are captured before grasping from two cameras, represented by Ib (**Figure 2**). Id is the position of the robot concerning the object grasped. Hence the vision feature fv can be calculated as

$$fv = R(Ib, Id) \tag{3}$$

where R represents the pre-trained neural network.

The images are passed through the standard convolutional network that uses the ResNet-18 architecture. Different from the previous work [22], the tactile sensors are used to obtain the force applied by the robot during the manipulation. As tactile sequences, the LSTMs are applied as the feature extractor:

$$ft = L(T0, T1, \dots, TT) \tag{4}$$

where ft is the last time step of the LSTMs' output and $T0, T1, \dots, TT$ is the input of the LSTMs at each step.

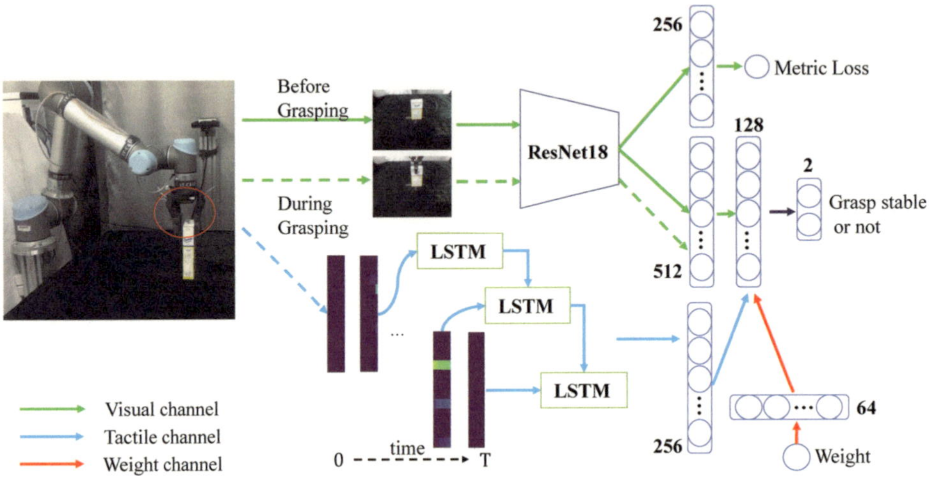

Figure 2.
Multimodal information predict grasp stability network.

Besides, the mass and mass distribution of the object also affect the stability of grasping. In order to simplify the problem, the weight of the object is known and the mass distribution is assumed uniform. Then the intrinsic object property is described as

$$fi = M(w) \tag{5}$$

where fi represents the intrinsic object feature and w is the object weight.

Then the multilayer perceptron (MLP) is used to extract the intrinsic feature. The sensory modalities provide the complementary information about the prospects for a successful grasp. For example, the camera's images show that the gripper is near the center of the object, and the tactile shows that the force is enough to keep stable for grasping. In order to study the method of multimodal fusion for predicting grasp outcomes, a neural network is trained to predict whether robot's grasp would be successful integrated by visual, tactile, and object's characters. The network computes $y = \{(X)\}$, where y is the probability of a successful grasp and $X = [fv, ft, fi]$ contains a set of images from multiple modalities: visual, tactile, and object intrinsic properties.

Train **the network**: initializing the weights of visual by CNN with a model pre-trained in Section III-A. The visual representation network is trained 200 epochs using the Adam optimizer [23], starting with a learning rate of 105 which is decreased by an order of magnitude halfway through the training process.

During the training, the RGB images are cropped with containing the table that holds the objects. Then, following the standard practice in object recognition, the images are resized to be 256 × 256 and randomly sampled at 224 × 224. Meanwhile the images are randomly flipped in the horizontal direction. However, the same data is still applied for augmentation to prevent overfitting.

3. Experiment and data collection

The experiment platform consists of the Eagle Shoal robot hand, two RealSense SR300 cameras, and the UR5 robot arm. As shown in **Figure 3**, they are arranged around the table of length 600 mm and width 600 mm. There is a layer of sponge on the surface of the table for protection. A soft flannel sheet covers the table to

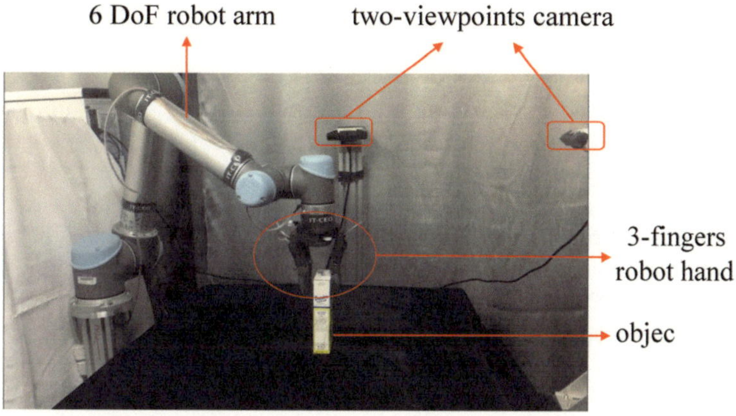

Figure 3.
Experiment platform.

Hand	Type	Weight	Force (mA)	Direction	Trail	Data type	Total
Eagle Shoal	10 objects	Empty	50/100/150	Top/right/back	10 times	T1/I	900 sets
Eagle Shoal	10 objects	Half/full	50/100/150	Top/right/back	10 times	T1/T2/I/V	1650 sets

Table 1.
Dataset statistics.

avoid the interference of light reflection. The UR5 robot arm with the Eagle Shoal robot hand fixes at the backside of the table. One RealSense camera is on the opposite side of the table for recording the front view of grasping. The other RealSense camera is located in the left of the table for recording the lateral view.

The general grasp dataset is built with various variables including shape, size, weight, grasp style, etc. The objects in the dataset contain different sizes of cuboid, cylinder, and special shapes, and their weights change by adding granules or water. Different grasping methods are tested by grasping from three directions including back, right, and top. The dataset with unstable grasping data is generated by slipping with added weight, changed grasp force, and adjusted grasp position (**Table 1**). The detailed processes are as follows:

1. The object is put on the center of the table; the front camera is used to get the point cloud data and computer the target's position.

2. Choose the object's half height position as the grasped point, control the robot to approach the object, and then add the random error of ±5 mm.

3. Based on the object's size, controlling the robot hand to grasp with a position loop mode, and then after 1 second, the robot arm lifts up with a speed of 20 mm/s.

4. After the robot arm moves to a certain position, the robotic finger's position is changed. If the hand is bending too much, this grasp is labeled as failure, and open the hand directly then prepare the next grasp.

5. The grasp is labeled as success, for a light object, if the robot puts down the object and, for some heavy object, the robot opens the hand and drops the object directly.

6. Putting the object on the center of the table, the robot arm returns to the initial place and waits for the next loop.

The proposed method is contrasted with traditional classifiers including k-nearest neighbor (KNN) [24], support-vector machine (SVM) [25], and naive Bayes (NB) [26]. A total of 2550 sets have been divided into 80% for training and 20% for testing. The KNN classifier is set with k = 3, and the SVM kernel is the radial basis function (RBF). The success rate with a criterion, the number of detection n and the number of label data m, is calculated by n/m. The contrast result in **Table 2** shows the performance of LSTM and SVM is both well with a success rate. However, the SVM's labels are on the falling edge, which means the SVM model gets a good classification result by learning the falling edge features. The falling edge means the object is dropped already and cannot help to realize a stable grasp. SVM proves unsuitable for this test.

Besides the success rate, another criterion is necessary to evaluate the slip detection. If the time of predict result turns from 1 to 0 ahead of the time in label data, set it as ahead sample and counted number n_{ahead}, calculate the ahead rate by n_{ahead}/m, and set it as the criterion. The results are shown in **Table 2**. With these two criteria, LSTM shows the superior performance that attains the higher success rate and higher ahead rate (**Figures 4** and 5).

Classifier	Success rate	Ahead drop	Ahead forecast
KNN	0.7970	0.8176	0.4961
SVM	0.8467	0.6667	0.2569
NB	0.6881	0.6569	0.4843
OUR	0.9460	0.8588	0.6373

Table 2.
Classification results of different classifiers.

Figure 4.
All the grasp object, from YCB object set.

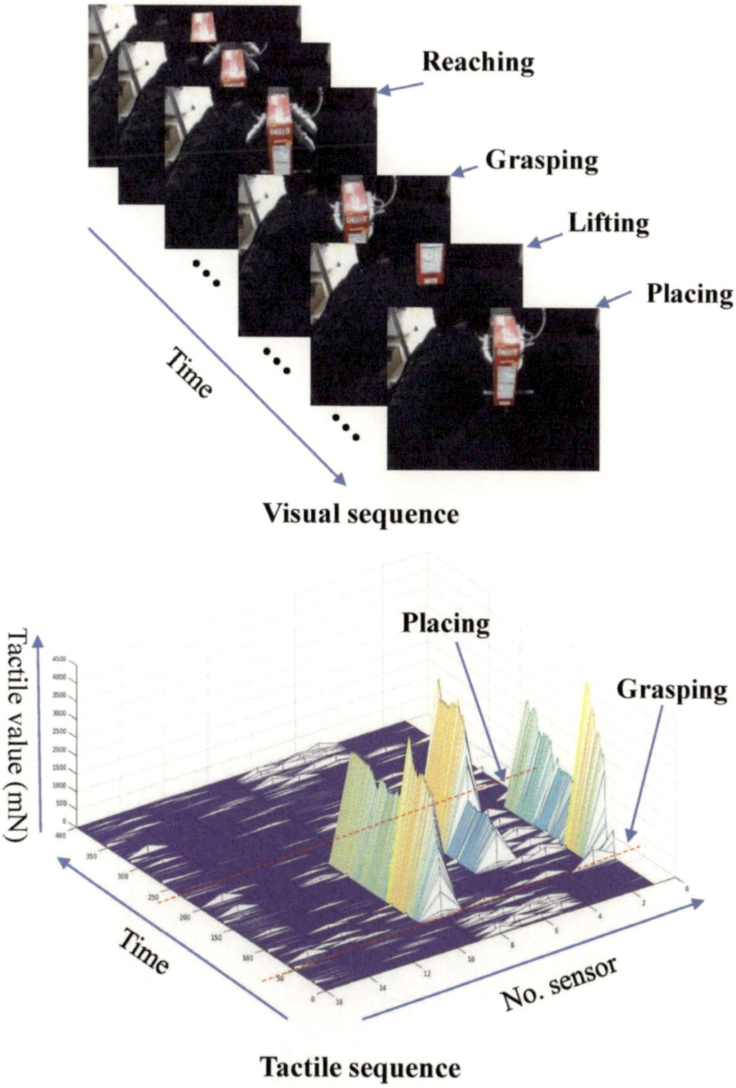

Figure 5.
Visual and tactile information visualization. Visual: grasping process video image sequence; and tactile: grasping process tactile sensor value.

4. Conclusions

In this chapter, the end-to-end approach for predicting stable grasp is proposed. Raw visual, tactile, and object intrinsic information are used, and the tactile sensor provides detailed information about contacts, forces, and compliance. More than 2500 grasp data are autonomously collected, and the multiple deep neural network model is proposed for predicting grasp stability with different modalities. The results show that visual-tactile fusion method improves the ability to predict grasp outcomes. In order to further validate the method, the real-world evaluations of the different models in the active grasp are implemented. Our experimental results demonstrate the superiority of the proposed method.

Acknowledgements

This work was supported by the Tsinghua University Initiative Scientific Research Program No. 2019Z08QCX15.

Author details

Bin Fang*, Chao Yang, Fuchun Sun and Huaping Liu
Department of Computer Science and Technology, Tsinghua National Laboratory for Information Science and Technology, Tsinghua University, Beijing, China

*Address all correspondence to: fangbin@tsinghua.edu.cn

References

[1] Bohg J, Morales A, Asfour T, Kragic D. Data-driven grasp synthesis survey. IEEE Transactions on Robotics. 2014;**30**(2):289-309

[2] Levine S, Pastor P, Krizhevsky A, Ibarz J, Quillen D. Learning hand-eye coordination for robotic grasping with deep learning and large-scale data collection. The International Journal of Robotics Research. 2018;**37**(4–5):421-436

[3] Guo D, Sun F, Kong T, Liu H. Deep vision networks for real-time robotic grasp detection. International Journal of Advanced Robotic Systems. 2016; **14**(1):1-8

[4] Mahler J, Pokorny FT, Hou B, Roderick M, Laskey M, Aubry M, et al. Dex-Net 1.0: A cloud-based network of 3D objects for robust grasp planning using a multi-armed bandit model with correlated rewards. In: 2016 IEEE International Conference on Robotics and Automation (ICRA). IEEE; 2016. pp. 1957-1964

[5] Mahler J, Liang J, Niyaz S, Laskey M, Doan R, Liu X, et al. Dex-net 2.0: Deep learning to plan robust grasps with synthetic point clouds and analytic grasp metrics. arXiv preprint arXiv: 1703.09312. 2017:1-12

[6] Mahler J, Matl M, Liu X, Li A, Gealy D, Goldberg K. Dexnet 3.0: Computing robust robot vacuum suction grasp targets in point clouds using a new analytic model and deep learning. arXiv preprint arXiv: 1709.06670. 2017:1-16

[7] Zhang T, McCarthy Z, Jowl O, Lee D, Chen X, Goldberg K, et al. Deep imitation learning for complex manipulation tasks from virtual reality teleoperation. In: 2018 IEEE International Conference on Robotics and Automation (ICRA). IEEE; 2018. pp. 1-8

[8] Vlack K, Mizota T, Kawakami N, Kamiyama K, Kajimoto H, Tachi S. Gelforce: A vision-based traction field computer interface. In: CHI'05 Extended Abstracts on Human Factors in Computing Systems. ACM; 2005. pp. 1154-1155

[9] Goldfeder C, Ciocarlie M, Dang H, Allen PK. The Columbia grasp database. In: IEEE International Conference on Robotics and Automation, 2009. ICRA'09. IEEE; 2009. pp. 1710-1716

[10] Sermanet P, Lynch C, Chebotar Y, Hsu J, Jang E, Schaal S, et al. Time-contrastive networks: Self-supervised learning from video. arXiv preprint arXiv:1704.06888. 2017:1-15

[11] Pinto L, Gupta A. Supersizing self-supervision: Learning to grasp from 50k tries and 700 robot hours. In: 2016 IEEE International Conference on Robotics and Automation (ICRA). IEEE; 2016. pp. 3406-3413

[12] Li J, Dong S, Adelson E. Slip detection with combined tactile and visual information. arXiv preprint arXiv:1802.10153. 2018:1-6

[13] Kobayashi F, Kojima F, Nakamoto H, Kida Y, Imamura N, Shirasawa H. Slip detection with multi-axis force/torque sensor in universal robot hand. International Journal of Applied Electromagnetics and Mechanics. 2012;**39**(1–4):1047-1054

[14] Heyneman B, Cutkosky MR. Slip classification for dynamic tactile array sensors. The International Journal of Robotics Research. 2016;**35**(4):404-421

[15] Chebotar Y, Hausman K, Su Z, Molchanov A, Kroemer O, Sukhatme G, et al. BiGS: BioTac grasp stability dataset. In: ICRA 2016 Workshop on Grasping and Manipulation Datasets; 2016

[16] Chebotar Y, Hausman K, Su Z, Sukhatme GS, Schaal S. Self-supervised regrasping using spatio-temporal tactile features and reinforcement learning. In: 2016 IEEE/RSJ International Conference on Intelligent Robots and Systems (IROS). IEEE; 2016. pp. 1960-1966

[17] Stachowsky M, Hummel T, Moussa M, Abdullah HA. A slip detection and correction strategy for precision robot grasping. IEEE/ASME Transactions on Mechatronics. 2016; **21**(5):2214-2226

[18] Kasper A, Xue Z, Dillmann R. The kit object models database: An object model database for object recognition, localization and manipulation in service robotics. The International Journal of Robotics Research. 2012;**31**(8):927-934

[19] Calli B, Walsman A, Singh A, Srinivasa S, Abbeel P. Benchmarking in manipulation research using the Yale-CMU-Berkeley object and model set. Robotics & Automation Magazine, IEEE. 2015;**22**(3):36-52

[20] Chechik G, Sharma V, Shalit U, Bengio S. Large scale online learning of image similarity through ranking. Journal of Machine Learning Research. 2010;**11**(Mar):1109-1135

[21] Chechik G, Sharma V, Shalit U, Bengio S. Large scale online learning of image similarity through ranking. Journal of Machine Learning Research. 2010;**11**:1109-1135

[22] Calandra R, Owens A, Upadhyaya M, Yuan W, Lin J, Adelson EH, et al. The feeling of success: Does touch sensing help predict grasp outcomes? arXiv preprint arXiv: 1710.05512. 2017:1-10

[23] Kingma DP, Ba J. Adam: A method for stochastic optimization. Journal of Computer Science. 2014:1-15

[24] Altman NS. An introduction to kernel and nearest-neighbor nonparametric regression. The American Statistician. 1992;**46**(3): 175-185

[25] Hearst MA, Dumais ST, Osuna E, Platt J, Scholkopf B. Support vector machines. IEEE Intelligent Systems and Their applications. 1998;**13**(4):18-28

[26] John GH, Langley P. Estimating continuous distributions in Bayesian classifiers. In: Proceedings of the Eleventh Conference on Uncertainty in Artificial Intelligence. Morgan Kaufmann Publishers Inc.; 1995. pp. 338-345

Chapter 8

Deep Learning-Based Detection of Pipes in Industrial Environments

Edmundo Guerra, Jordi Palacin, Zhuping Wang and Antoni Grau

Abstract

Robust perception is generally produced through complex multimodal perception pipelines, but these kinds of methods are unsuitable for autonomous UAV deployment, given the restriction found on the platforms. This chapter describes developments and experimental results produced to develop new deep learning (DL) solutions for industrial perception problems. An earlier solution combining camera, LiDAR, GPS, and IMU sensors to produce high rate, accurate, robust detection, and positioning of pipes in industrial environments is to be replaced by a single camera computationally lightweight convolutional neural network (CNN) perception technique. In order to develop DL solutions, large image datasets with ground truth labels are required, so the previous multimodal technique is modified to be used to capture and label datasets. The labeling method developed automatically computes the labels when possible for the images captured with the UAV platform. To validate the automated dataset generator, a dataset is produced and used to train a lightweight AlexNet-based full convolutional network (FCN). To produce a comparison point, a weakened version of the multimodal approach—without using prior data—is evaluated with the same DL-based metrics.

Keywords: deep learning, autonomous robotics, UAV, multimodal perception, computer vision

1. Introduction

Robotics, as a commercial technology, started to be widespread some decades ago, but instead of decreasing, it has been growing year by year with new contributions in all the related fields that it integrates. The introduction of new materials, sensors, actuators, software, communications and use scenarios converted Robotics in a pushing area that embraces our everyday life. New robotic morphologies are the most shocking aspect that society perceives (i.e., the first models of each type generally produce the largest impact), but the long-term success of robotics is found in its capability to automate productive processes. Manufacturers and developers know that the market is found not only in large-scale companies (car manufacturers and electronics mainly) but also in the SME that provides solutions to problems that are manually performed so far. Also, robotics has opened the doors to new applications that did not exist some years ago and are also attractive to investors. These facts, together with lower prices for equipment, better programming and communication tools, and new fast-growing user-friendly collaborative robotic frameworks, have pushed robotics technology at the edge in many areas.

It is clear that industrial robotics leads the market worldwide, but social/gaming uses of robots have increased sales. Nevertheless, the most promising scenario for the present time and short term is the use of robots in commercial applications out of the plant floor. Emergency systems, inspection, and maintenance of facilities of any kind, rescues, surveillance, agriculture, fishing, border patrolling, and many other applications (without military use) attract users/clients because their use increases the productivity of the different sectors, low prices and high profitability are the keys.

There exist many robot morphologies and types (surface, underwater, aerial, underground, legged, wheels, caterpillar, etc.) but authors want to draw attention in the unmanned aerial vehicles (UAVs), which have several properties that make them attractive for a set of application that cannot be done with any other type of robot. First, those autonomous robots can fly, and therefore, they can reach areas that humans or other robots cannot. They are light, easy to move from one area to another, and can be adapted to any area, terrain, soil, building, or facility. The drawback is the fragility in front of adverse meteorological events, and their autonomy is quite limited compared with unmanned surface vehicles (USVs).

UAVs have seen the birth of a new era of unthinkable cheap, easy applications up to now. The authors would like to focus its use in the maintenance and inspection of industrial facilities, but specifically in the inspection of pipes in big, complex factories (mainly gas and oil companies) where the manual inspection (and even location and mapping) of pipes becomes an impossible task. Manned helicopters (with thermal engines) cannot fly close to pipes or even among a bunch of pipes. Scaffolds cannot be put up in complex, unstable, and fragile pipes to manually inspect them. Therefore, a complex problem can be solved through the use of UAVs for inspecting pipes of different diameters, colors, textures, and conditions in hazardous factories. This problem is not new and some solutions have been brought to an incipient market. Works as those in [1, 2] propose the creation of a map of the pipe set navigating among it with odometry and inertial units [3]. Obstacle avoidance in a crowded 3D world of pipes becomes of great interest when planning a flight; in [4], some contributions are made in this direction although the accuracy of object is deficient to be a reliable technology. Work in [5] overcomes some of the latter problems with the use of a big range of sensors, cameras, laser, barometer, ultrasound, and a computationally inefficient software scheme made the UAV too heavy and unreliable due to the excessive sensor fusion approach.

Many of the technical developments that have helped robotics grow have had a wider impact, especially those related with increasing computational power and parallelization levels. Faster processors, with tens of cores and additional multiple threat capabilities, and modern GPUs (graphics processing unit) have led to the emergence of GPGPU (general-purpose computing on GPU). These type of computing techniques have led to huge advances in the artificial intelligence (AI) field, producing the emergence of the "deep learning" field. The deep learning (DL) field is focused in using artificial neural networks (ANNs) that present tens or hundreds of layers, exploiting the huge parallelization capabilities of modern GPU. This is used in exploiting computational cores (e.g., CUDA cores), which compared on a one-to-one basis with a processor core, they are less powerful and slower, but can be found in amounts of hundreds or thousands. This has allowed the transition from shallow ANN to the deeper architectures and innovations such as several types of convolutional layers. In this work, the authors present a novel approach to detect pipes in industrial environments based in fully convolutional networks (FCNs). These will be used to extract the apparent contour of the pipes, replacing most of the architecture developed in [6] and discussed in Section 2. To properly train these networks, a custom dataset relevant to the domain is required, so the authors captured a dataset and developed an automatic label generation procedure base in previous works. Two different state-of-the-art semantic segmentation approaches were trained and evaluated with the standard metrics to prove the validity of the whole approach. Thus, in the following section, some generalities about the pipe detection and positioning problem are discussed, and the authors' previous work [6] on it, as it will be relevant later. The next section discusses the semantic segmentation problem as a way to extract the apparent contour, both surveying classical methods, considered for earlier works, and state of the art deep-learning-based methodologies. The fourth section describes how the automatic label generator using multimodal data was derived and some features to the process. The experimental section starts discussing the metrics employed to validate the results, the particularities of the domain dataset generated and describes how an AlexNet FCN architecture was trained through transfer learning and the results achieved. To conclude, some discussion on the quality of the results and possible enhancements is introduced, discussing which would be the best strategies to follow continuing this research.

2. Related work

As it has been discussed, inspection and surveying are a frequent problem where UAV technologies are applied. The most common scenario found is that of a hard to reach infrastructure that is visually inspected through different sensors onboard a piloted UAV. Some projects have proposed the introduction of higher level perception and automation capacities, depending on the specific problem. In these cases, it is common to join state-of-the-art academic and industrial expertise to reach functional solutions.

In one of these projects, the specific challenge of accurately detecting and positioning a pipe in real time using only the hardware deployable in a small (per industry standards) UAV platform was considered (**Figure 1**), with several solutions studied and tested (including vision- and LIDAR-based techniques).

In the case of LIDAR-based detection, finding a pipe is generally treated as a segmentation problem in the sensor space (using R3 data collected as "*point clouds*"). There are many methods used for LIDAR detection, but the most successful are based on stochastic model fitting and registration, commonly in RANSAC (Random Sample Consensus [7]) or derived approaches [8, 9]. Three different data density levels were tested using the libraries available through ROS: using RANSAC over a map estimated by a SLAM technique, namely LOAM [10]; detecting the pipe

Figure 1.
One of the UAV used for the development of perception tasks in the AEROARMS project. Several sensors were deployed, processing them with a set of SBCs (single-board computers), including a Velodyne LiDAR, two different cameras, ultrasonic range-finder (height), and optical flow.

in a small window of consecutive point clouds joined by an ICP-like approach [11]; and finally to simply work using the most recent point cloud. The first approach probed to be computationally unfeasible, no matter what optimization was tested, as even working with a single datum cloud point could be prohibitive if not done carefully. To enhance the performance, the single cloud point approach was optimized employing spatial and stochastic filtering to reduce the data magnitude, and a curvature filter allowed to reduce fake positives in degenerate configurations, producing robust results at between 1 and 4 Hz. To solve the same problem with visual sensors, a two-step strategy was used. In order to estimate the pose of the pipes to be found, they were assumed to be circular and regular enough to be modeled as a straight homogeneous circular cylinder. This allowed using a closed-form conic equation [12], which related the axis of the pipe (its position and orientation as denoted in Plücker coordinates) with the edges of its projection in the image space. While this solves the positioning problem, the detection probed to be a little more challenging: techniques based on edge detection, segmentation, or other classical computer vision methods used to work under controlled light but failed to perform acceptably in outdoor scenarios. This issue was solved by introducing human supervision, where an initial seed for the pipe in the image sensor space was provided (or validated) by a human and then tracked robustly through vision predicting it with the UAV odometry.

With these results, discussed in [6], it was apparent that a new solution was needed, as the LiDAR approaches were too slow and the vision-based techniques probed themselves unreliable. The final proposed solution was based on integrating data from the laser and the vision sensors: the RANSAC over LiDAR approach would detect robustly the pipe and provide an initial position, which would then be projected into the image space (accounting for displacements if odometry is available) and used as a seed for the vision-based pipeline described.

In that same work [6], a sensibility analysis studying the effects of the relative pose between the sensor and pipes is provided. Once the pipe is detected in the LiDAR's space sensor, the cylinder model is projected into the R^2 image space using a projection matrix derived from the calibrated camera model (assumed to be a thin lens pinhole model, per classic literature [13]). This provides a region or band of interest where to look for the edges of the pipe in the image and is useful to solve the degenerate conic equation up to scale (i.e., being a function of the radius). An updated architecture version of the process is depicted in **Figure 2**.

The detailed architecture of the multimodal approach reveals how the LiDAR-based pipeline minimizes the data dimensionality by filtering non-curved surfaces (i.e., remove walls, floor, etc.) and also by removing entirely regions of the sensed space if priors or relevant data or the expected relative position of the pipe to the sensor is available. This was aimed at minimizing the size of the point cloud to be processed by the RANSAC step. To be able to project the detected pipe from the LiDAR sensor space into the camera image, some additional information was required: the rigid transformation between sensors (i.e., the calibration between LiDAR and camera) and an estimation of the odometry of the UAV. This is due because, even in the best assumption, with a performance slightly over 4 Hz, the delay between the captured point cloud and the produced estimation of the pipe would be over 200 ms. Therefore, the projection of the detected pipe to predict the area of interest to search the apparent contour has to consider the displacement during this period, not only the rigid LiDAR to camera transformation. This predicted region of interest is used in the vision process pipeline, with predictions of the appearance of the pipes into image space used to refine the contour search. This contour search relies on stacking a Hough transform to join line segment detector (LSD) detected segments (to overcome partial obstructions) on the relevant area

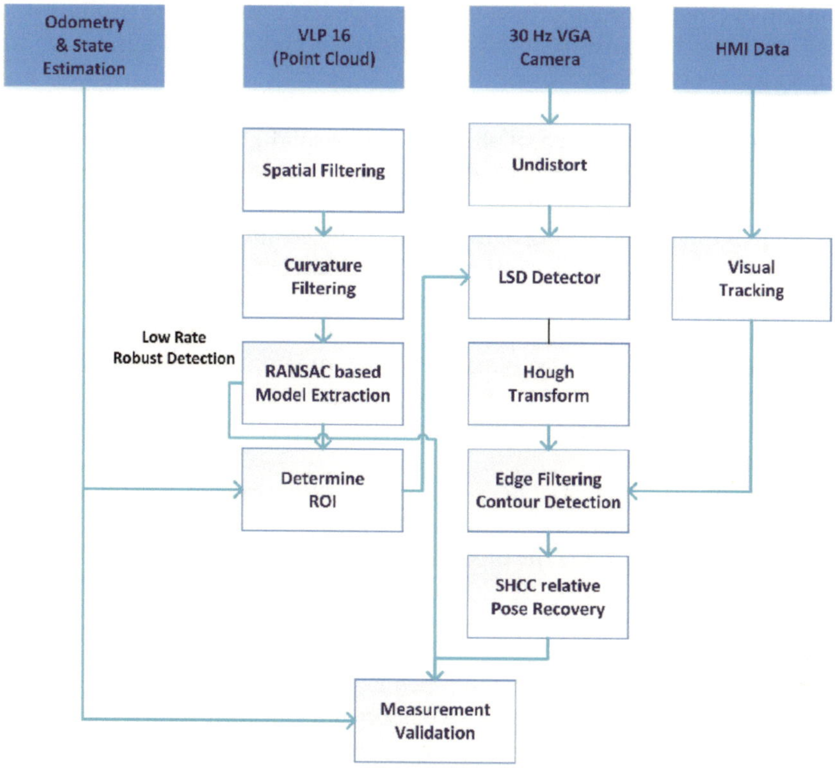

Figure 2.
The architecture of the multimodal perception pipeline combining LiDAR and camera vision. An updated version adds to previous works a validation step using odometric measurements.

and allows to choose the nearest correctly aligned lines. Notice that using a visual servoing library [14], an option to use data provided through human interaction was kept as available, though the integration of LiDAR detections as seeds into the visual pipeline made it unnecessary. To avoid degenerate or spurious solutions, a validation step (based on reprojection and "matching" of the Plückerian coordinates [15] for a tracked piped) was later introduced.

This architecture leads to a fast (limited by the performance of the vision-based part) and robust (based on the RANSAC resilience to spurious detections) pipe detector with great accuracy, which was deployed and test in a UAV. The main issue of the approach is the hardware requirements: access to odometry from the avionics systems, LiDAR, and camera sensors, and enough computing power to process them (beyond any other task required from the UAV). All this hardware is focused on solving what can be described as a semantic segmentation problem. This is relevant given the enormous changes produced in the last decade in the computer vision field, and how classic problems like semantic segmentation are currently solved.

3. Semantic segmentation problem: classic approaches and deep learning (DL)

In the context of computer vision, the semantic segmentation problem is used to determine which regions of an image present an object of a given category, that is, a class or label is assigned to a given area (be it a pixel, window, or segmented region). The different granularity accepted is produced by how the technique and its solution evolved: for a long time, it was completely unfeasible to produce pixel-wise solutions, so images were split according to different procedures, which added a complexity layer to the problem.

Current off-the-shelf technologies have changed the paradigm, as GPUs present huge capabilities in terms of parallelization, while solid-state disks make fast reliable storage cheap. These technical advancements have increased dramatically the performance, complexity, and memory available for data representation, especially for techniques inherently strong in highly parallelized environments. One of the fields where the impact has been more noticeable has been the artificial intelligence community, where the artificial neural network (ANN) has seen a resurgence thanks to the support this kind of hardware provides to otherwise computationally unfeasible techniques. The most impactful development in recent years has been the convolutional neural networks (CNNs), which have become the most popular computer vision approach for several of the classic problem and the default solution for semantic segmentation.

To understand the impact of deep learning into our proposed solution, we will discuss briefly how the classical segmentation pipeline worked and how the modern CNN-based classifier became the modern semantic segmentation techniques.

3.1 Classic semantic segmentation pipeline

The classic semantic segmentation pipeline can be split into two generic blocks, namely image processing for feature extraction and feature level classification. The first block generally includes any image preprocessing done and resizing/resampling, splitting the image into the regions/windows, defining the granularity level of the classification, and finally, extracting the features itself. The features can be of any type and frequently the ones feed to the classification modules will be a composition of several individual features from different detectors. The use of different window/region-based approaches helps build up higher level features, and the classification can be refined at later stages with data from adjacent regions.

Notice that this kind of architecture generally relies on classifiers which required very accurate knowledge or a dataset with the classes to learn specified for each input so it can be trained. **Figure 3** shows the detection of pipelines in classic semantic segmentation. Notice that to train the classifier, the image mask or classification result becomes also an input for the training process.

So, it can be seen that solving the semantic segmentation problem though classic pattern recognition methods requires acute insight into the specifics of the problem domain, as the features to be detected and extracted are built/designed specifically. This implies (as mentioned earlier) working from low-level features and explicitly deriving the higher level features from them is a very complex problem itself, as they are affected by the input characteristics, what is to be found/discriminated, and which techniques will be used in the classification part of the pipeline.

3.2 The segmentation problem with deep learning

Modern semantic segmentation techniques have organically evolved with the rise of the deep learning field to its current prominence. This evolution can be seen as a refinement in the scale of the inference produced from very coarse (image level probabilistic detection) to very fine (pixel level classification). The earliest ANN examples made probabilistic predictions about the presence of an object of a given class, that is, detection of objects with a probability assigned. The next step, achieved thanks to increased parallelization and network depth, was starting to tackle the localization problem, providing centroids and/or boxes for the detected

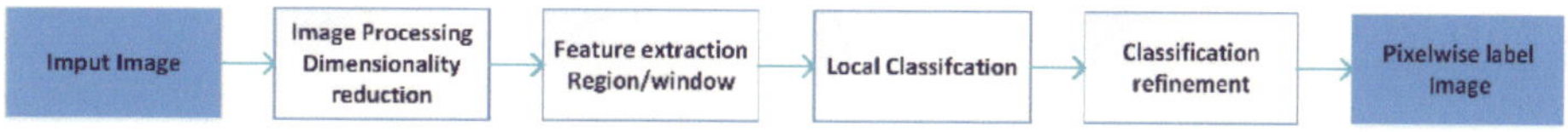

Figure 3.
Block diagram of a classical architecture approach for semantic segmentation using computer vision.

classes (the use of *classes* instead of *objects* here is deliberate, as the instance segmentation problem, separating adjacent objects of the same class, would be dealt with much later).

The first big break into the classification problem was done by AlexNet [13] in 2012, when it won the ILSVRC challenge, with a score of 84.6% in the top-5 accuracy test, while the next best score was only 73.8% (based on classic techniques). AlexNet has since then become a known standard and a default network architecture to test problems, as it is actually not very deep or complex (see **Figure 4**). It presents five convolutional layers, with max-pooling after the first two, three fully connected layers, and a ReLU to deal with non-linearities. This clear victory of the CNN-based approaches was validated next year by Oxford's VGG16 [16], one of the several architectures presented, winning the ILSVRC challenge with a 92.7% score.

While several other networks have been presented with deeper architecture, relevant development focused on introducing new types of structures into the networks. GoogLeNet [17], the 2014 ILSVRC winner, achieved victory thanks to the novel contribution of the inception module, which validated the concept that the CNN layers of a network could operate in other orders different from the classic sequential approach. Another relevant contribution produced by technology giants was ResNet [18], which scored a win for Microsoft in 2016. The introduction of residual blocks allowed them to increase the depth to 152 layers while keeping initial data meaningful for training the deeper layers. These residual blocks architecture essentially forwards a copy of the received inputs of a layer; thus, later layers received the results and same inputs of prior layers and can learn from the residuals.

More recently, ReNet [19] architecture was used to extend recurrent neural networks (RNNs) to multidimensional inputs.

The jump from the classification problem with some spatial data to pixel level labeling (refining inference from image/region to pixel level) was presented by Long [20], with the fully convolutional network (FCN). The method they proposed was based on using the full classifier (like the ones just discussed) as layers in a convolutional network architecture. FCN architecture, and its derivatives like U-Net [21] are the best solutions to semantic segmentation for most domains. These derivatives may include classic methods, such as DeepLab's [22] conditional random fields [23], which reinforces the inference from spatially distant dependencies, usually lost due to CNN spatial invariance. The latest promising contributions to the semantic segmentation problem are based on the encoder-decoder architecture, known as autoenconders, like for example SegNet [24].

For the works discussed in this chapter, a FCN16 model with AlexNet as a semantic segmentation model was used. The main innovation introduced by the general FCN was exploiting the classification power via convolution of the common semantic segmentation DL network, but at the same time, reversing the downsampling effect of the convolution operation itself. Taking AlexNet as an example, as seen in **Figure 4**, convolutional layers apply a filter like operation while reducing the size of the data forwarded to the next layer. This process allows producing more accurate "deep features" but at the same time also removes high-level information describing the spatial relation between the features found. Thus, in order to exploit the features from the deep layers while the keeping information from spatial relation, data from multiple layers has to be fused (with element-wise summation).

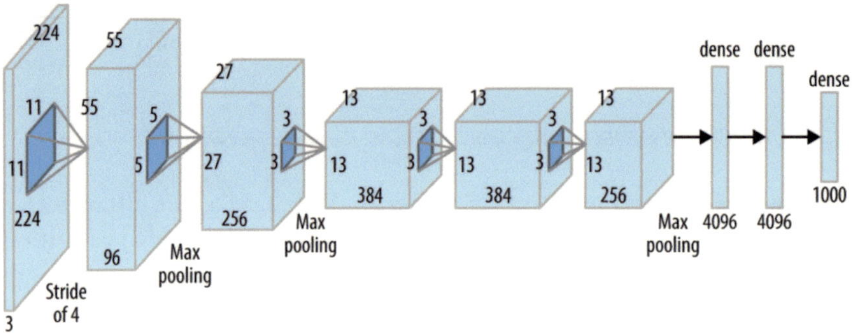

Figure 4.
Diagram of the AlexNet architecture, showcasing its pioneering use of convolutional layers.

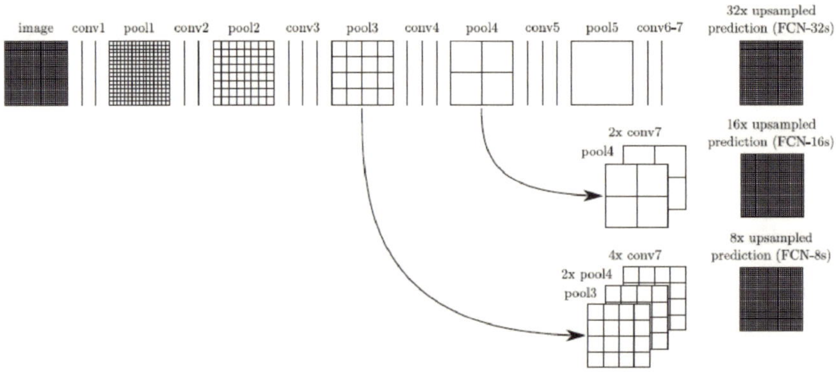

Figure 5.
Detail of the skip architectures (FCN32, FCN16, and FCN8) used to produce results with data from several layers to recover both deep features and spatial information from shallow layers (courtesy of [25]).

In order to be able to produce this fusion, data from the deeper layers are upsampled using deconvolution. Notice that data from shallow layers will be coarser but contain more spatial information. Thus, up to three different levels can be processed through FCN, depending on the quantity of layers deconvoluted and fused, as seen in **Figure 5**.

More information on the detailed working of the different FCN models can be found in [25]. It is still worth noting that the more shallow layers are fused, the more accurate the model becomes, but according to the literature, the gain from FCN16 to FCN8 is minimal (below 2%).

4. Automated ground truth labeling for multimodal UAV perception dataset

Classic methods using trained classifiers would pick designed features (based on several metrics and detectors, as discussed earlier) to parametrize a given sample and assign a label. This would allow creating small specific datasets, which could be used to infer the knowledge to create bigger datasets in a posterior step. The high specificity of the features chosen (generally with expert domain knowledge applied implicitly) with respect to the task generally made them unsuitable to export learning to other domains.

By contrast, deep learning offers several transfer learning options. That is, as it was proven by Yosinski [26], trained with a distant domain dataset are generally

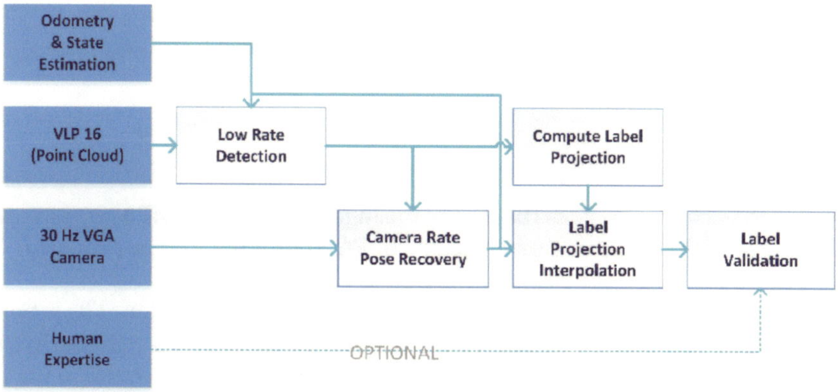

Figure 6.
The framework proposed to automatically produce labeled datasets with the multimodal perception UAV.

useful for different domains and usually better than training from an initial random state. Notice that the transferability of features decreases with the difference between the previously trained task and the target one and implies that the network architecture is the same up to the transferred layers at least.

With this concept in mind, we decided to build a dataset to train an outdoor industrial pipe detector with pixel level annotation to be able to determine the position of the pipe. While the ability of transfer learning allows us to skip building a dataset with several tens of thousands of images, and therefore, the authors will work with a few thousand, which were used to fine-tune the network. These orders of magnitude are required as a "shallow" deep network. For instance, the AlexNet already presents 60 million parameters.

Capturing and labeling a dataset is a cumbersome task, so we also set to automatize this task with minimal human supervision/interaction, exploiting the capabilities of the sensing architecture proposed in earlier works described in Section 2.

This framework, see **Figure 6**, uses the images captured by the UAV camera sensor, the data processed by the localization approach chosen (see Section 2) to obtain the UAV odometry, and pipe detection seeds from the RANSAC technique treating the LiDAR point cloud data. When a pipe (or generally a cylinder) is detected and segmented in the data sensor provided by the LiDAR, this is used to produce a label for the temporally near images, to identify the region of the image (the set of pixels) containing the pipe or cylinder detected and its pose w.r.t. the camera. Notice that even running the perception part, the camera works at a higher rate than the LiDAR, so the full odometric estimation is used to interpolate between pipe detections, to estimate where the label should be projected into the in-between images (just as it was described for the pipe prediction in Section 2).

This methodology was used to create an initial labeled dataset with actual data captured in real industrial scenarios during test and development flights, as it will be discussed in the next section.

5. Experimental evaluation

To evaluate the viability of the proposed automated dataset generation methodology, we apply it to capture a dataset and train several semantic segmentation networks with it. To provide some quantitative quality measurement for the solutions produced, we use modified standard metrics for state-of-the-art deep learning, accounting that in our problem we are dealing with only one semantic class:

- **PA** (pixel accuracy): base metric, defined by the ratio between properly classified pixels *TP* and the total number of pixels in an image, pix_{total}:

$$PA = \frac{TP}{pix_{total}} \tag{1}$$

Notice that usually, besides the PA the mean pixel accuracy (MPA) is provided, but in our case, it reduces to the same value of PA, thus it will not be provided.

- **IoU** (intersection over union): standard metric in segmentation. The ration is computed between the intersection and union of two sets, namely the found segmentation and the labeled ground truth. Conceptually, it equals to the ratio between the number of correct positives (i.e., the intersection of the sets) TP, over all the correct positives, spurious positives FP and false negatives FN (i.e., the union of both the ground truth and segmentation provided). Usually, it is used as mean IoU (MIU), averaging the same ratio for all classes.

$$IoU = \frac{TP}{TP + FP + FN} \tag{2}$$

An additional metric usually computed along with the MIU is the frequency weighted MIU, which just weighs the average IoU computed at MIU according to the relative frequency of each class. The MIU, in our case, IoU is the most relevant metric and the most widely used when reporting segmentation results (semantic or otherwise).

5.1 Dataset generation results

The system proposed was implemented over the ROS meta-operating system, just as in previous works [6], where the UAV system used to capture the data is described. A set of real flights in simulated industry environments was performed, where flights around a pipe were done. During these flights, averaging ~240 s, an OTS USB camera was used to capture images (at 640 × 480 resolution), achieving an average frame rate of around 17 fps. This translated in around 20,000 raw images captured, including the parts of flight where no industry-like elements are present, thus of limited use.

Notice that as per the method described, the pipe to be found can be only labeled automatically when the LiDAR sensor can detect it; thus, the number of images was further reduced due to the range limitations of the LiDAR scanner. Other factors, such as vibrations and disruptions in the input or results of required perceptual data, further reduced the number of images with accurate labels.

Around ~2100 images were automatically labeled with a mask assigning a ground truth for the pipe in the image. After an initial human inspection of the assigned label, a further ~320 were rejected, obtaining a final set of 1750. The image rejected produced spurious ground truths/masks. Some of them had inconsistent data and the reprojection of the cylinder detected in through RANSAC in LiDAR scans was not properly aligned (error could be produced by spurious interpolation of poses, faulty synchronization data from the sensors, or due to deformation of the UAV frame, as it is impossible for it to be perfectly rigid). Another group presented partial detections (only one of the edges of the pipe is visible in the image), thus making it useless for the apparent contour optimization. A third type of error found was produced by the vision-based pipeline, where a spurious mask was generated, commonly some shadows/textures displace/retort the edge, or areas not pertaining to the pipe are assigned due similarity of the texture and complexity of delimiting the areas.

A sample of the labeling process can be seen in **Figure 7**, with the original image, the segmented pipe image, and approximations to centroid and bounding bow.

Out of the several options available to test the validity of the dataset produced, the shallow architecture AlexNet was selected, as it could be easily trained and it would provide some insight in the performance that could be realistically expected from a CNN-based approach deployed in the limited hardware of a UAV.

According to previous literature, the dataset was divided into training, validation, and test at the standard ratio of 70, 15, and 15% respectively.

To match the input of AlexNet the images were resized to 256 × 256 resolution. This was mainly done to reduce the computational load, as the input size could be easily fit adjusting some parameters, like the stride. To train and test the network, the Pytorch library was used, which provides full support for its own implementation of AlexNet.

To produce some metrics relevant to the network architecture just trained, a modified version of the technique used to label the dataset was used. Note that this approach, as described in previous sections, uses LiDAR, cameras, and odometry to: acquire an initial robust detection (from LiDAR), track its projection and predict it in the camera image space (using odometric data), and finally determine its edges/contour in the image. The robustness of the LiDAR detection is mainly due to exploiting prior knowledge (in the form of the known radius of the pipe to detect) that cannot be introduced into the AlexNet architecture to produce a meaningful comparison. So, a modified method, referred to as NPMD (no-priors multimodal detector) was employed to estimate the accuracy of earlier work detector without priors. The main difference was modifying the LiDAR pipeline to be able to detect several pipes with different radius (as it should be considered unknown). This led to the appearance of false positives and spurious measurements, which in turn weakened the results produced by the segmentation part of the visual pipeline.

Thus, FCN with AlexNet classification was trained using a pre-trained model for AlexNet, with the standard stochastic gradient descend (SGD) with a momentum of 0.9. A learning rate of 10^{-3} was used, according to known literature, with image batches of 20. The weight decay and bias learning rate were set to standard values of 5.10^{-4} and 2, respectively. Without any prior data, and no benefit to obtain by doing otherwise reported in any previous works, the classifier layer was set to 0, and the dropout layer in the AlexNet left unmodified. This trained model produced the results found in **Table 1**.

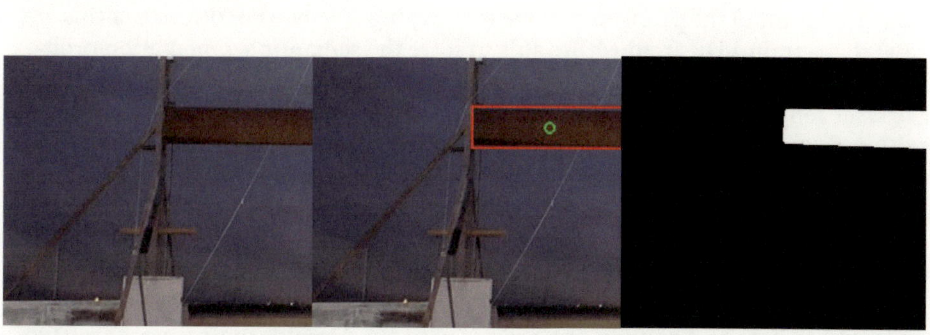

Figure 7.
Left: dataset image. Middle: bounding box and centroid of the region detected. Right: segmentation mask image.

	AlexNetFCN	UPMD
PA	73.4	56.7
IoU	58.6	42.1

Table 1.
Experimental results obtained by AlexNet-based FCN.

It can be seen that eliminating the seed/prior data from the multimodal detector made it rather weak, with very low values for IoU, signaling the presence of spurious detections and probably fake positives. The FCN-based solution was around 1.5 times better segmenting the pipe, being a clear winner. This was to be expected as we deliberately removed one of the key factors contributing to the LiDAR-based RANSAC detection robustness, the radius priors, leading to the appearance of spurious detections.

It is worth noting that although the results are not that strong in terms of metrics achieved for a single-class case, there are no other vision-only pipe detectors with better results in the literature, neither other approaches actually tested in real UAV's platforms, like authors' previous works [6].

6. Conclusions

The field of computer vision has been greatly impacted by the advances in deep learning that have emerged in the last decade. This has allowed solving, with purely vision-based approaches, some problems that were considered unsolvable under this restriction. In the case presented, a detection and positioning problem, solved with limited hardware resources (onboard a UAV) in an industry-like uncontrolled scenario through a multimodal approach, has been solved with a vision-only approach. The previous multimodal approach relied in LiDAR, cameras, and odometric measurements (mainly from GPS and IMU) to extract data with complex algorithms like RANSAC and combine them to predict the position of a pipe and produce a measurement. This system was notable thanks to its robustness and performance but presented the huge requirements detailed in [6]. In order to solve the problem in a simpler and more affordable manner, a pure visual solution was chosen as the way to go, exploring the deep learning opportunities.

Although the switch to a pure visual solution meant that during its use, the procedure would only use the camera sensor, the multimodal approach was still used to capture data, and through a series of modifications, turn it into an automatic labeling tool. This allowed building a small but complete dataset with fully labeled images relevant to the problem that we were trying to solve. Finally, to test this dataset, we train a DL architecture able to solve the semantic segmentation problem. Thus, three different contributions have been discussed in this chapter: firstly, a dataset generator exploiting multimodal data captured by the perception system to be replaced has been designed and implemented; secondly, with this dataset generation tool, the data captured has been properly labeled so it can be used for DL applications; and finally, a sample lightweight network model for semantic segmentation, FCN with AlexNet classification, has been trained and evaluated to test the problem.

By the same reasons that there was no dataset available for our challenge and we had to capture and develop one dedicated to our domain, there were no related works to obtain metrics. In order to have some relevant metrics to compare the results of the developed approach, a modified version of [13] was produced and benchmarked without the use of prior knowledge. Under these assumptions, the new CNN-based method was able to clearly surpass the multimodal approach, though it still lacks robustness to be considered ready for industrial standards. Still, these initial tests have proven the viability of the built dataset generator and the utilization of CNN-based semantic segmentation to replace the multimodal approach.

Acknowledgements

This research was funded by the Spanish Ministry of Economy, Industry and Competitiveness through Project 2016-78957-R.

Author details

Edmundo Guerra[1], Jordi Palacin[2], Zhuping Wang[3] and Antoni Grau[1*]

1 Automatic Control Department, Technical University of Catalonia UPC, Barcelona, Spain

2 Department of Informatics and Industrial Engineering, University of Lleida - UdL, Lleida, Spain

3 Department of Control Science and Engineering, Tongji University, Shanghai, China

*Address all correspondence to: antoni.grau@upc.edu

References

[1] Wang Z, Zhao H, Tao W, Tang Y. A new structured-laser-based system for measuring the 3D inner-contour of pipe figure components. Russian Journal of Nondestructive Testing. 2007;**43**(6):414-422

[2] Song H, Ge K, Qu D, Wu H, Yang J. Design of in-pipe robot based on inertial positioning and visual detection. Advances in Mechanical Engineering. 2016;**8**(9):168781401666767

[3] Hansen P, Alismail H, Rander P, Browning B. Visual mapping for natural gas pipe inspection. The International Journal of Robotics Research. 2015;**34**(4-5):532-558

[4] Zsedrovits T, Zarandy A, Vanek B, Peni T, Bokor J, Roska T. Visual detection and implementation aspects of a UAV see and avoid system. In: IEEE. 2011. pp. 472-475. [cited 05 March 2018].Available from: http://ieeexplore.ieee.org/document/6043389/

[5] Holz D, Nieuwenhuisen M, Droeschel D, Schreiber M, Behnke S. Towards multimodal omnidirectional obstacle detection for autonomous unmanned aerial vehicles. ISPRS—International Archives of the Photogrammetry, Remote Sensing and Spatial Information Sciences. 2013;**1**:201-206

[6] Guerra E, Munguía R, Grau A. UAV visual and laser sensors fusion for detection and positioning in industrial applications. Sensors. 2018;**18**(7):2071

[7] Fischler MA, Bolles RC. Random sample consensus: A paradigm for model fitting with applications to image analysis and automated cartography. Communications of the ACM. 1981;**24**(6):381-395

[8] Choi S, Kim T, Yu W. Performance evaluation of RANSAC family. Journal of Computer Vision. 1997;**24**(3):271-300

[9] Raguram R, Frahm J-M, Pollefeys M. A comparative analysis of RANSAC techniques leading to adaptive real-time random sample consensus. In: ECCV 2008 (Lecture Notes in Computer Science). Berlin, Heidelberg: Springer; 2008. pp. 500-513

[10] Zhang J, Singh S. LOAM: Lidar odometry and mapping in real-time. In: Proceedings of the "Robotics: Science and Systems. 2014" conference, July 12-16, 2014, Berkeley, USA. 2014. pp. 1-9

[11] Besl PJ, McKay ND. A method for registration of 3-D shapes. IEEE Transactions on Pattern Analysis and Machine Intelligence. 1992;**14**(2):239-256

[12] Doignon C, de Mathelin M. A degenerate conic-based method for a direct fitting and 3-d pose of cylinders with a single perspective view. In: Proceedings 2007 IEEE International Conference on Robotics and Automation. Roma, Italy: IEEE; 10-14 July 2007. pp. 4220-4225

[13] Krizhevsky A, Sutskever I, Hinton GE. Imagenet classification with deep convolutional neural networks. In: Pereira F, Burges CJC, Bottou L, Weinberger KQ, editors. Advances in Neural Information Processing Systems 25. Curran Associates, Inc.; 2012. pp. 1097-1105

[14] Marchand E, Spindler F, Chaumette F. ViSP for visual servoing: A generic software platform with a wide class of robot control skills. IEEE Robotics and Automation Magazine. 2005;**12**(4):40-52

[15] Bartoli A, Sturm P. The 3D line motion matrix and alignment of line reconstructions. In: Proceedings of the 2001 IEEE Computer Society Conference on Computer Vision and Pattern Recognition, CVPR, Kauai, HI,

USA. Vol. 1. 8-14 December 2001. pp. I-287-I-292

[16] Simonyan K, Zisserman A. Very Deep Convolutional Networks for Large-Scale Image Recognition. 2015. Available from: http://arxiv.org/abs/1409.1556

[17] Szegedy C, Liu W, Jia Y, Sermanet P, Reed S, Anguelov D, et al. Going deeper with convolutions. In: 2015 IEEE Conference on Computer Vision and Pattern Recognition (CVPR). Boston, MA, USA; 7-12 June 2015. pp. 1-9

[18] He K, Zhang X, Ren S, Sun J. Deep residual learning for image recognition. In: 2016 IEEE Conference on Computer Vision and Pattern Recognition (CVPR). Las Vegas, NV, USA; 27-30 June 2016. pp. 770-778

[19] Visin F, Kastner K, Cho K, Matteucci M, Courville A, Bengio Y. ReNet: A Recurrent Neural Network Based Alternative to Convolutional Networks. 2015. Available from: http://arxiv.org/abs/1505.00393

[20] Long J, Shelhamer E, Darrell T. Fully convolutional networks for semantic segmentation. In: 2015 IEEE Conference on Computer Vision and Pattern Recognition (CVPR). Boston, MA, USA; 7-12 June 2015. pp. 3431-3440

[21] Ronneberger O, Fischer P, Brox T. U-Net: Convolutional networks for biomedical image segmentation. In: Navab N, Hornegger J, Wells WM, Frangi AF, editors. Medical Image Computing and Computer-Assisted Intervention—MICCAI 2015. (Lecture Notes in Computer Science). Cham: Springer International Publishing; 2015. pp. 234-241

[22] Chen L-C, Papandreou G, Kokkinos I, Murphy K, Yuille AL. DeepLab: Semantic image segmentation with deep convolutional nets, Atrous convolution, and fully connected CRFs. IEEE Transactions on Pattern Analysis and Machine Intelligence. 2018;**40**(4):834-848

[23] Lafferty J, Mccallum A, Pereira F. Conditional random fields: Probabilistic models for segmenting and labeling sequence data. In: 2001 Conference on Machine Learning, ICML, Williamstown, MA, USA; 28 June-1 July 2001. pp. 282-289

[24] Badrinarayanan V, Kendall A, Cipolla R. SegNet: A deep convolutional encoder-decoder architecture for image segmentation. IEEE Transactions on Pattern Analysis and Machine Intelligence. 2017;**39**(12):2481-2495

[25] Shelhamer E, Long J, Darrell T. Fully convolutional networks for semantic segmentation. IEEE Transactions on Pattern Analysis and Machine Intelligence. 2017;**39**(4):640-651

[26] Yosinski J, Clune J, Bengio Y, Lipson H. How transferable are features in deep neural networks? In: Advances in Neural Information Processing Systems. Montrea (CANADA): Curran Associates, Inc.; 2014. pp. 3320-3328

Chapter 9

Visual-Inertial Indoor Navigation Systems and Algorithms for UAV Inspection Vehicles

Lorenzo Galtarossa, Luca Francesco Navilli and Marcello Chiaberge

Abstract

In UAV navigation, one of the challenges in which considerable efforts are being focused is to be able to move indoors. Completing this challenge would imply being able to respond to a series of industrial market needs such as the inspection of internal environments for safety purpose or the inventory of stored material. Usually GPS is used for navigation, but in a closed or underground environment, its signal is almost never available. As a consequence, to achieve the goal and ensure that the UAV is able to accurately estimate its position and orientation without the usage of GPS, an alternative navigation system based on visual-inertial algorithms and the SLAM will be proposed using data fusion techniques. In addition to the navigation system, we propose an obstacle avoidance method based on a Lidar sensor that allows navigation even in the absence of light.

Keywords: unmanned aerial vehicle (UAV), GPS denied, indoor navigation, Lidar, inertial measurement unit (IMU), visual-inertial odometry (VIO), 3D reconstruction

1. Introduction

Nowadays indoor navigation is one of the most popular topics in the field of scientific research; this is because it is itself a very interesting challenge from the scientific point of view and with a potentially huge impact on the current market with multiple applications ranging from the industrial sector to the relief sector in emergency situations.

As well known, while in outdoor navigation, UAVs use Global Position System (GPS) signal to easily understand, with good accuracy, their position in space, in indoor navigation the possibility to use this technology for positioning and localization decays [1–3]. In fact, the Global Navigation Satellite System (GNSS) in indoor environments is almost blocked or made too weak by buildings, walls or several potential sources of interference.

The first approaches to indoor UAV navigation used technologies such as Wi-Fi, Bluetooth or Ultra-Wide Band [4]. The disadvantage and the great limitation of these technologies is that they need to structure or prepare the environment in order to be able to navigate inside with the UAVs.

In recent years, the miniaturization of both visual sensors and computers that guarantee good computing power and at the same time less weight has made possible a new different approach to the topic of UAV indoor navigation.

This approach is based on inertial and visual systems, for example, see [5–7], with enormous advantage of being free from any kind of need to structure the environment and therefore potentially flexible and universal. The position of the UAV is estimated using inertial measurements provided by gyroscopes and accelerometers that are now available in every smartphone and with small dimensions and weights. The accuracy of this type of inertial measurements is good but not sufficient in order to guarantee a precise indoor positioning. In fact the estimate of the UAV position based only on inertial systems tends to diverge and drift over time due to the fact that inertial measurement unit (IMU) measurements are corrupted by noise and bias with the results in pose estimates unreliable for long-term navigation. To avoid the effects of this phenomenon, the inertial system is combined with a visual one that uses a camera to collect information and extract futures from the surrounding environment and track them over time in order to estimate the trajectory of the camera. This approach is usually referred to as visual-inertial odometry (VIO). Information from the camera can also be used to build a map of the environment and then perform what is called simultaneous localization and mapping (SLAM).

In this chapter we propose a system architecture that allows a UAV to inspect a tunnel, which is a closed environment, navigating autonomously. To estimate the position of the UAV in the absence of GPS, we used Robust Visual Inertial Odometry (ROVIO) which is a predictor of inertial visual states based on an extended Kalman filter (EKF) that combines the visual information of a monocular camera with the measurements derived from the IMU inertial platform. At the same time, a navigation and obstacle avoidance algorithm based purely on a Lidar sensor is proposed.

The UAV is equipped with a companion computer in which Robotic Operating System (ROS) is installed and allows the processing of information coming from the monocular camera and the IMU as well as those coming from the Lidar for the navigation.

Furthermore, a scheduling system has been implemented and embedded on the computer companion that allows to set different strategies to approach the inspection of the tunnel before starting the mission. Defined safety patterns that are activated in case of dangerous situations for UAV and humans are also into the scheduling system.

The chapter is organized as follows. In Section 2, we briefly analyze and describe the sensors used and their characteristics, and we go into detail on how the architecture of the system is defined. In Section 3, we explain which criteria characterize the navigation system and what is the logic behind it. In conclusion we present the results achieved, outlining the performance of the proposed system for indoor navigation evaluating possible improvements for future research.

2. System architecture

This chapter describes the overall system architecture under different points of view. We start with a short description of ROS, that is, the framework that allows to manage different UAV's operation. Then we move to analyze the hardware and payload of the UAV, we describe all the crucial characteristics and we explain why those characteristics and components are crucial for the project. Afterwards we explain why between the several VIO algorithms implemented in the past years, we select and use ROVIO and how we design the visual-inertial sensor. Moreover, we propose a scheduling system based on some setting parameters that are crucial to well set up at the beginning of the mission in order to define the positioning of the UAV inside the tunnel. These parameters allow to define both inspection and navigation settings, the last ones are useful to change based on the geometrical characteristic of the tunnel. Instead, the parameters define the type of inspection that is needed to be performed in order to collect data or achieve the inspection's objectives.

2.1 Robot operating system

The heart of the whole system is robot operating system (ROS); it is an open source framework to manage robot's operations, tasks and motions. Among the several features that ROS has, the most relevant is the availability of code, packages and open source projects. This is a key element in the development of complex systems which often encompass different skills and concepts [8, 9].

A set of processes can be represented in a graph as a node that can receive, send and process messages, called topics, coming from other sensors, actuators and nodes.

In this system the two main topics for the construction of the algorithm are those of the Lidar and the odometry that give to the system the information about the obstacles around the drone (coming from the Lidar) and the pose outgoing from ROVIO which defines the position and the orientations of the UAV along the six DOF.

The information on these two messages is fed to the navigation algorithm which returns the topic of the speed to be assigned to the drone during the inspection.

2.2 UAV's payload

Referring to **Figures 1** and **2**, the main components of the UAV are:

1. Custom frame with a propulsion system designed for a total payload of about 4 kg

2. LED lighting system for navigation and acquisition of frames even in complete darkness and absence of light

3. Cameras for the acquisition of photograms that allow the construction of a three-dimensional model of the inspected tunnel and environment

4. Visual-inertial sensor used for positioning, control and as the main source of odometry

5. Laser sensors for detecting distance from the ground

6. Voltage and current distribution system, mainly 12 and 5 V

7. LiDAR 2D laser scanner for detecting obstacles and relative distances

As a payload there is also a mini computer companion that has the necessary power to perform, record, process and analyze data from all the sensors and to move the UAV accordingly.

Figure 1.
Assembly UAV payload, first perspective.

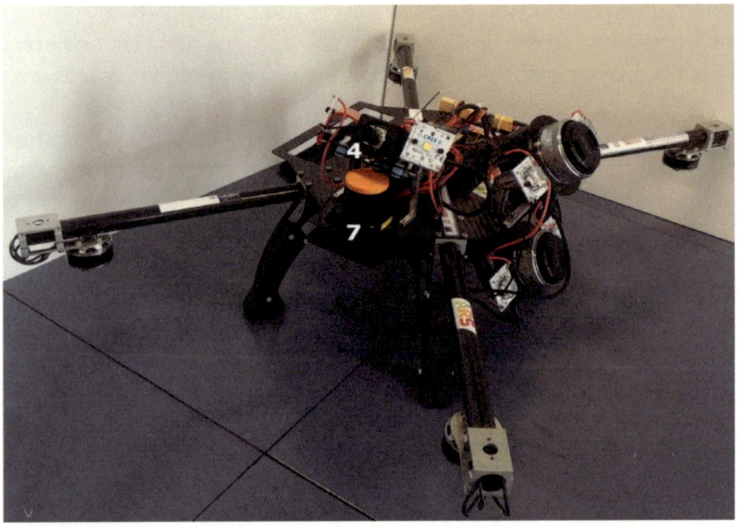

Figure 2.
Assembly UAV payload, second perspective.

2.3 Visual-inertial sensor

As mentioned in the introduction, the lack of a GPS requires the use of a sensor that can guarantee the correct positioning inside a closed space. In particular, follow the research trend in the field of computer vision; the sensor is composed by a monocular camera and an IMU inertial measurement sensor. These two sensors are connected to each other by a mechanism called hardware trigger. This choice was made to ensure maximum precision in the acquisition of data from both sensors since it is a crucial point in order to obtain a precise positioning of the UAV. The kind of sensors described above is preferable to purely visual-based techniques or any other sensor configurations for large number of advantages:

- Unlike monocular simultaneous localization and mapping (SLAM) based only on visual sensor, the generated maps have an absolute scale.

- Status estimation and feature tracking, which allow to understand how the UAV is moving in the space, are more robust to the motion blur and fast rotations than exclusive visual-based system.

- IMU data can be used to provide instantaneous estimates at over 100 Hz.
- The installation of this hardware is cheaper, smaller, lighter and lower in power consumption with respect to the three-dimensional laser scanner or even stereo configurations.

However, this type of approach has two main problems: the first one is related to the IMU camera timestamp synchronization that can cause large errors and drift in the state estimate of the UAV. The second one is that the system has to be able to continuously estimate and compensate the drift and the distortions of the IMU data. These problems are mainly related to the VIO algorithm chosen for this project, ROVIO [10]. ROVIO is a visual-inertial state estimator based on EFK which proposed several novelties. In addition to FAST corner features, whose 3D positions are parameterized with robot-centric bearing vectors and distances, multi-level patches are extracted from image stream around these features. These patch features are tracked and warped based on IMU predicted motion, and the photometric errors are used in the update step as innovation terms. The choice to use ROVIO is made based on the average CPU load of the visual-inertial algorithms proposed by [6]; in fact, the CPU usage—considering the limited CPU resources of the computer companion and the amount of all the operations to be performed during the UAV mission—was considered the main aspect on which to base the overall system design.

ROVIO, unlike other odometry systems (e.g. mono VINS) that attempt to compensate the time's errors, requires that all timestamps be accurate in order to work properly. Considering this aspect, several manual experiments have been done in order to investigate the incidence of the time synchronization and timestamp acquisition on ROVIO. From our experiments we can see that the temporal accuracy depends on both application and the state estimator, but more generally we can say that the range of time acquisition must be between 2 and 5 milliseconds. Besides this threshold, it is no longer possible to follow rapid movements that cause the divergence and drift of the overall system. On the other side, below two milliseconds, we do not perceive huge improvements from the operational point of view.

Most of the camera sensors acquire their timestamp when the image is sent to the computer companion. However, there are many potential sources of delay that can affect the accuracy of the timestamp related to an image like the exposure time of the camera, the internal data processing, internal filter (from IMU point of view), data transfer and also the OS scheduler of the camera. For most camera sensors on the market, these delays are generally including between 5 and 30 milliseconds. While some delays related to the exposure or some other parameters of the camera are constant or can be expected, unknown delays prevent, to the computer companion point of view, from providing accurate timing information to any visual-inertial estimator.

For this reason, we decide to use a custom-made sensor directly linked to a microcontroller that receives data from the IMU and use a trigger line to check when the camera captures images. When an image is taken, as consequence, the microcontroller transmits information about the timestamp and IMU to the computer companion that links it to the image coming from the visual sensor. **Figure 3** shows the schematic of the circuit between microcontroller, IMU, camera and computer companion, while **Figure 4** the two visual and inertial sensors.

2.4 Scheduling system

The overall system, **Figure 5**, is designed as follow: through a web application linked to a web server, the user can select and set the parameters of the mission.

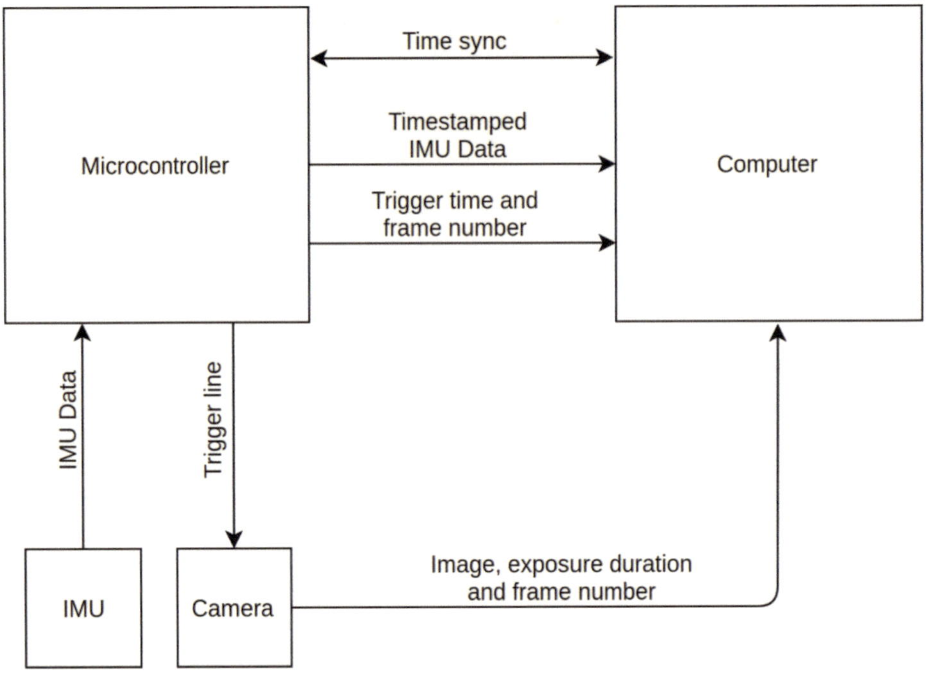

Figure 3.
How computer companion, microcontroller and sensors are linked.

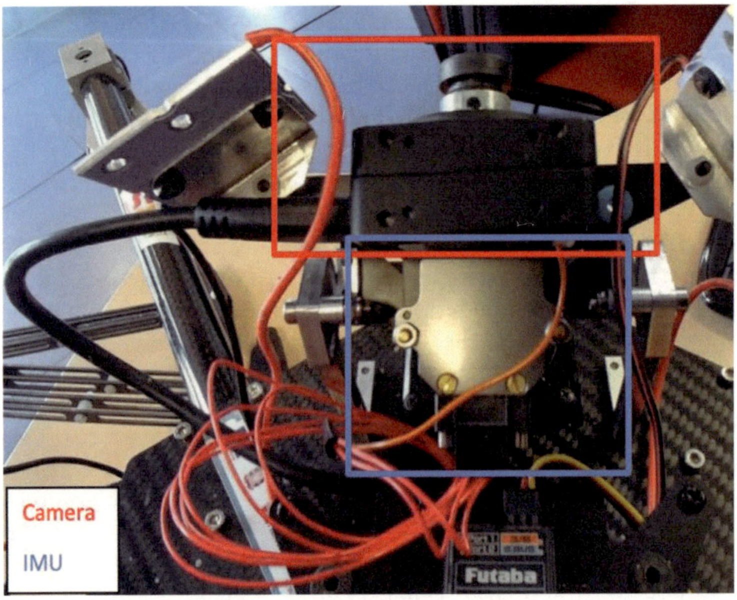

Figure 4.
Camera and IMU sensors.

These parameters are loaded through the scheduling system in which some patterns related to the overall status check of the UAV system (e.g. battery status, sensors status, LED status) are implemented.

At the lower level, there are the ROS nodes responsible for navigation, VIO and flight controller manager that execute the commands translated by the scheduling system.

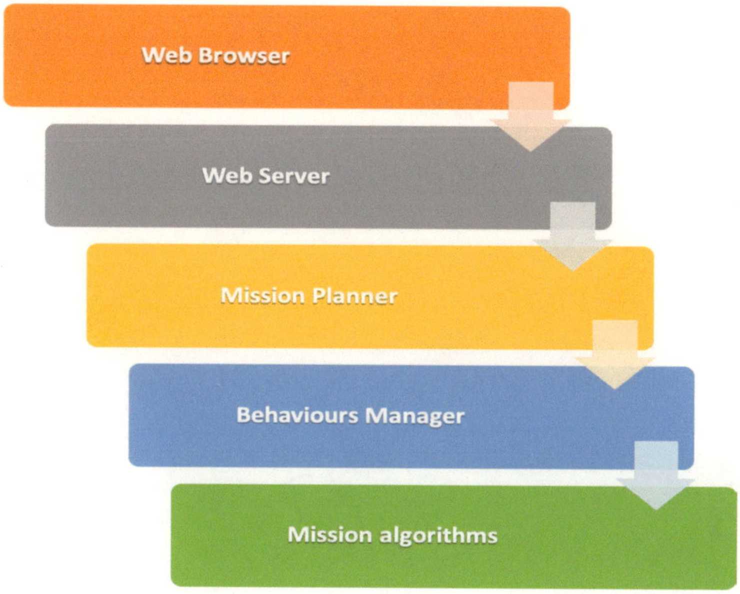

Figure 5.
Logic behind the system.

Parameter	Type of mission
Tunnel diameter (m)	1/2
Distance to travel (m)	2
Position altitude (m)	1/2
Data record: camera and Lidar (on/off)	1/2
Cruise speed (m/s)	1/2
K positioning (K)	1/2
Come back (on/off)	1/2
Maximum distance (m)	1/2
Minimum distance (m)	1/2

Table 1.
Setting parameters for each type of mission.

The scheduling system is based on SMACH (http://wiki.ros.org/smach) that is a task-level architecture based on ROS for rapidly creating complex robot behaviour. In this application the possible behaviors are two and depend from the type of mission that the user selected through the web GUI:

- Mission 1: Complete exploration of the tunnel

- Mission 2: Partial exploration of the tunnel (fixed distance chosen from the user)

For both missions it is possible to specify if the UAV must return to the home position or land at the end of the tunnel once the exploration is completed. Moreover, there are some specifications that the user can select by GUI. These parameters are related to the geometry of the tunnel and some working condition and are obviously related to the type of mission selected (**Table 1**).

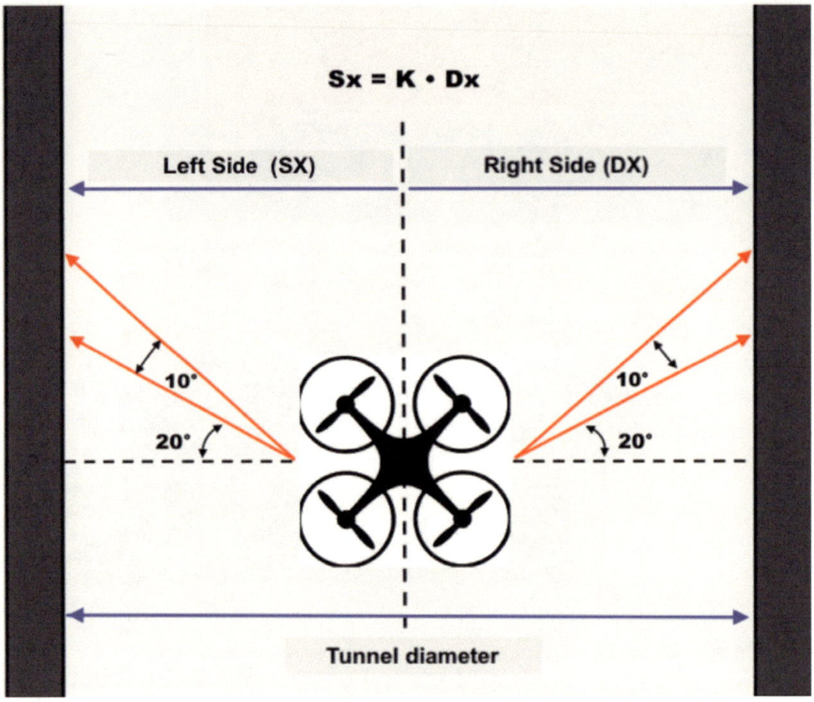

Figure 6.
K parameter logic.

The K parameter indicates the position that the UAV must maintain, while the tunnel inspection is defined as the ratio between the distance of the UAV with the right and left walls of the tunnel (**Figure 6**).

In the hypothesis in which K has been defined equal to 1, the drone will carry out the mission remaining in a central position with respect to the left and right walls. In the same way, with K = 2, the distance held to the left wall by the UAV will be doubled compared to the right distance.

The same ratio is maintained even during return to home navigation, when the reference system of the drone will be rotated 180° on the *xy* plane.

This positioning system was thus implemented to allow a 3D reconstruction of the tunnel inspected by using a single camera.

3. Flight system

Navigation and obstacle avoidance are one of the fundamental problems in mobile robotics, which are being already studied and analyzed by the researchers in the past 40 years. The goal of navigation is to find an optimal path from a starting to the goal point with obstacle avoidance competence. In order to guarantee an autonomous navigation, the robot must be able to safeguard a certain reliability in terms of position (IMU, GPS or other sensors) and ensure a map sufficiently precise to generate a path without collisions and faithful to the real one.

When the robot is in a complete unknown area and does not have information about the surrounding area, the global motion planning fails and does not produce any solution [11]. For this kind of situations, the local motion planning is more suitable.

The objective of the obstacle avoidance is to move a robot towards an area that is free of collisions thanks to the information handled by the sensors during the motion execution, which are steadily updated [12].

In this chapter the autonomous flight system will be defined. In particular, all the aspects concerning the navigation and the related intrinsic logic will be explained.

3.1 Navigation algorithm

Given the application context, a blind tunnel of semi-circular or circular cross-section with a diameter ranging from 2 to 5 metres, it was necessary to develop a specific navigation algorithm that would allow the UAV to explore the surrounding environment avoiding obstacles that could arise during the investigation of the tunnel. The environment taken into consideration for the definition of the algorithm was structured in a tunnel with an entrance and an exit, where there were no bifurcations of the channel.

Within a dark and unknown environment, the use of a Lidar is crucial to carry out navigation in an appropriate manner and for the implementation of the obstacle avoidance algorithm.

Light detection and ranging (Lidar) is a remote sensing technique that allows to determine the distance of an object or a surface using a laser pulse. The distance of the object is determined by measuring the time elapsed between the pulse emission and the reception of the retro-diffused signal. In the same way, to define the height from the ground, the height sensor is necessary. It allows stabilization of the UAV and its navigation to a predefined altitude with the possibility, thanks to the autopilot, of enabling terrain following or the technology that in an automatic way maintains a constant relative distance with respect to the ground.

The main task of the Lidar sensor is to monitor three distances during the navigations. The three distances are one front to the drone navigation and the two laterals, considering a 20° of inclination with respect to the perpendicular drone (**Figure** 7).

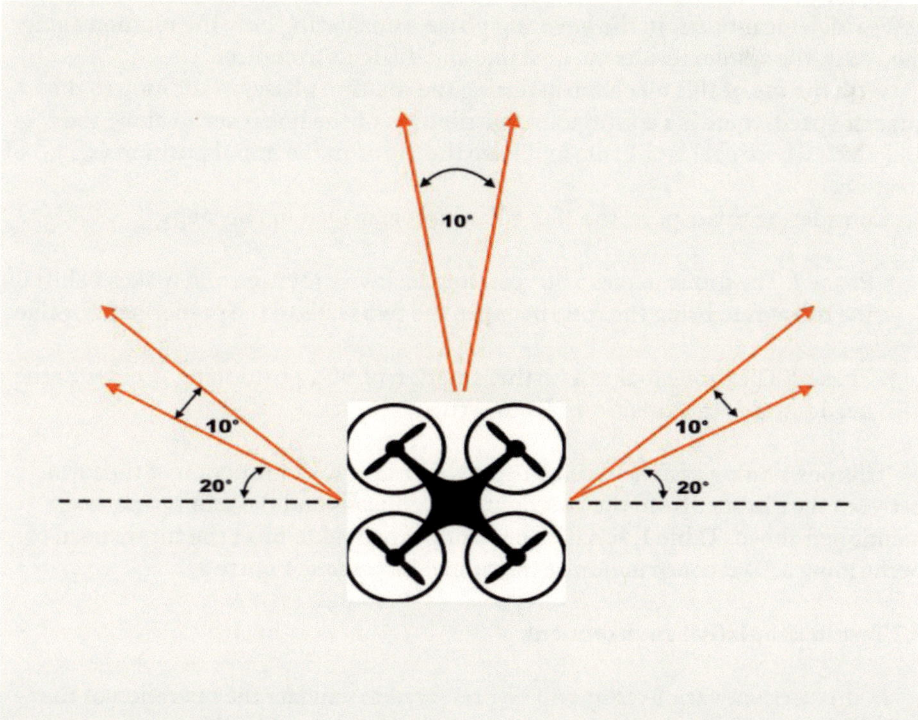

Figure 7.
Monitoring of the three distances.

This solution allows to monitor the frontal space to make sure that the path is free from obstacle, while the two lateral distances serve to guarantee the correct positioning within the UAV tunnel. Being that a single acquisition in any direction may not return completely valid information (optical sensor readings may be subject to disturbance and error depending on the type of surface, color and material on which the signal bounces), it is thought to acquire more data for each direction in a range of 10° in order to achieve a satisfactory level of consistency of the data.

The acquisition of the front distances is necessary to avoid hitting an obstacle present inside the tunnel and more importantly, once the tunnel is investigated in its entire length, recognize the end and be able to start the landing operation. The threshold set for the frontal control has been limited to 5 metres (maximum distance). This implies that until no obstacle is identified in this radius, the UAV will proceed to a predefined cruising speed (1 m/s); on the contrary, if an obstacle is detected, the speed will begin to decrease directly proportional to the distance between the UAV and the above obstacle.

At the minimum threshold value, 2 m from the obstruction, the drone resets its speed by stopping and remaining in hovering condition.

Recognizing the impossibility of advancing the UAV has two possible strategies to pursue: the first strategy involves the initialization of the landing operation, whereas the second includes first a 180° of rotation and then proceeding to the home positioning. Which of the two operations carried out is decided by the operator during the planning of the mission?

Another crucial point of the project was the planning of the rotations that had to be carried out when the anti-collision system recognized the end of the tunnel.

This phase was managed with the aid of the rotation matrices, with the aim of maintaining, during the rotation phase of 180°, the position saved immediately before the start of the rotation. This system had to be studied due to the problem brought by the vibrations and the imperfect balance of the payload installed on the UAV which meant that, in the hovering phase, considering only the rotation along the z axis, the system results are unstable and difficult to control.

With the use of this mechanism during the rotation phase, in addition to the angular speed, there is a continuous contribution of the linear speed along the x and y axis whose goal is to bring and keep the drone in the initial position (x_0, y_0) of rotation.

Considering this aspect, the 180° rotation is managed in two steps:

- Phase 1: The drone makes a 90° counterclockwise rotation and makes a shift on the roll axis to bring the ratio between the two walls to the predefined K value.

- Phase 2: The drone makes a further rotation of 90°, positioning itself with the head towards the direction of round trip.

This positioning system defined using a constant K as a function of the ratio between the distance from the two right and left walls and the other parameters mentioned above, **Table 1**, has thus been implemented to meet the future need of performing a 3D reconstruction of the tunnel inspected (**Figure 8**).

3.2 Test in simulative environment

In this section, various tests will be presented to validate the operation of the entire system and the obtained results, which are evaluated in different unknown indoor environments such as tunnels, to describe the advantages and limitations of this project.

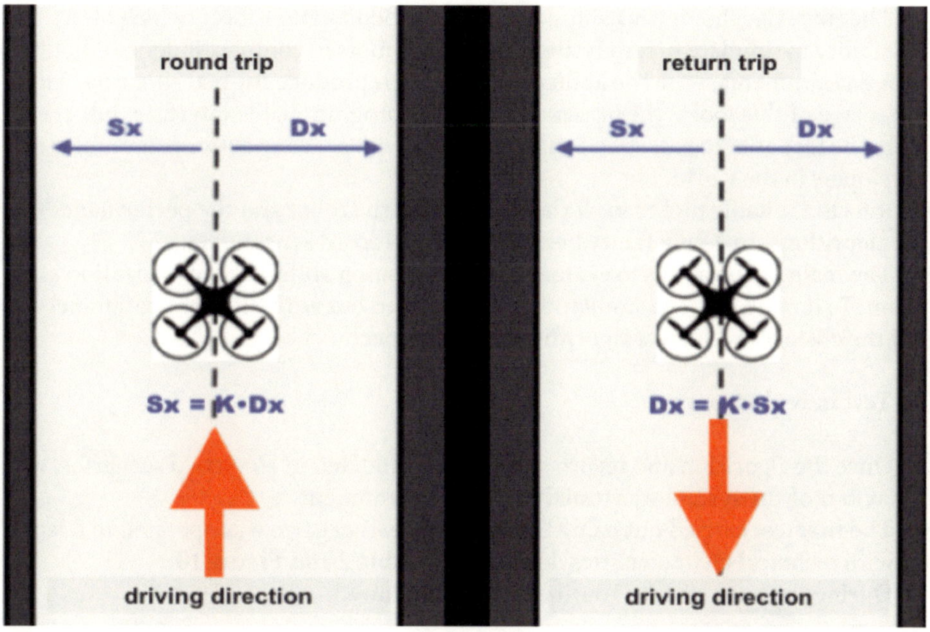

Figure 8.
Tunnel exploration.

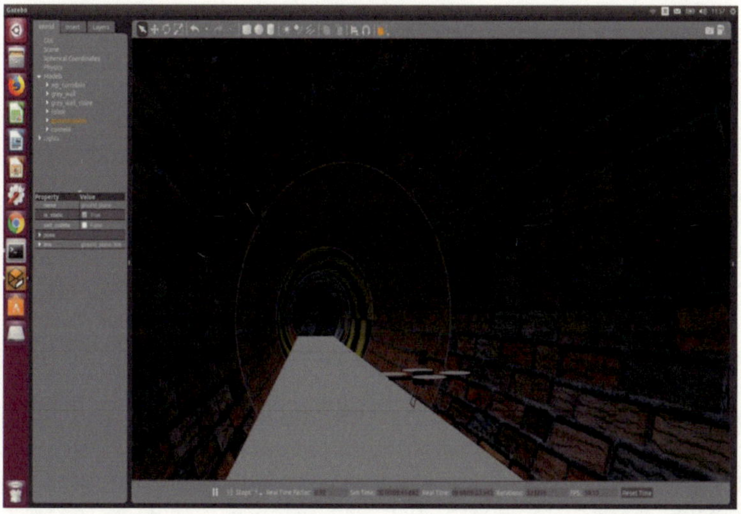

Figure 9.
UAV test in simulative environment.

In order to validate the navigation algorithm presented in this document, before performing real field tests, it was preferred to apply a more precautionary approach by testing the logic of the software in the simulation field.

This kind of approach is preferred for UAV since the failure of navigation frequently involves serious damage to the hardware and therefore, in cascade, a strong impact on the cost of the project.

To assess the quality of the software developed, the first tests were performed in a simulative environment using a UAV model (**Figure 9**). The simulative environment was defined using Gazebo, while Rviz was used to display the results (both tools are provided by ROS).

The first, Gazebo, is a 3D simulator for rigid bodies and robots, which offers the possibility to simulate precisely and efficiently robots in complex indoor and outdoor environments, with the ability to faithfully reproduce the real situation. The advantage of this tool is the presence of an easy programmable interface, but even more the fact of being an open source software with a strong active community of developers in the world.

Rviz is a suitable tool to view the 3D status of the robot and the performance of the algorithms, to debug faulty behaviors and to record sensor data.

The main purpose was to evaluate the functioning ability of the navigation algorithm. To do this, various simulations were carried out with different parameters, to test the obstacle avoidance algorithm in every aspect.

3.3 Test in real scenario

Once the algorithm and its procedure were validated in all virtual scenarios, the behaviour of the system was tested in a real environment.

The first test carried out using the drone in real scenario was operated in a facility with technical characteristics described in **Table 2** and **Figure 10**.

During the test a precise routine has been followed:

1. UAV positioning at the beginning of the tunnel.

2. System power-on and lipo-battery connection on UAV.

3. Check communication link between UAV and ground station.

4. Execution of ROS launch file.

5. Set up mission parameters.

6. Start mission.

The types of tests that have been performed are divided into two categories:

- Type A: tests conducted in a lighting environment

- Type B: tests with on board LED lighting, in a dark environment

Table 3 shows the results obtained for type B condition and the relative absolute error calculated as the difference in Euclidean distance traveled by the UAV between the point of take-off and point of landing. The distances over which the tests were performed are respectively 10, 20 and 30 m, iterated 10 times in order to compare the error related to the odometry data. **Table 4** displays the average minimum and maximum error for each different test.

Stretch	Height	Width	Length
a	2.15	2.40	11
b	2.10	2.35	30
c	1.90	2.40	5

Table 2.
Characteristics of the tunnel.

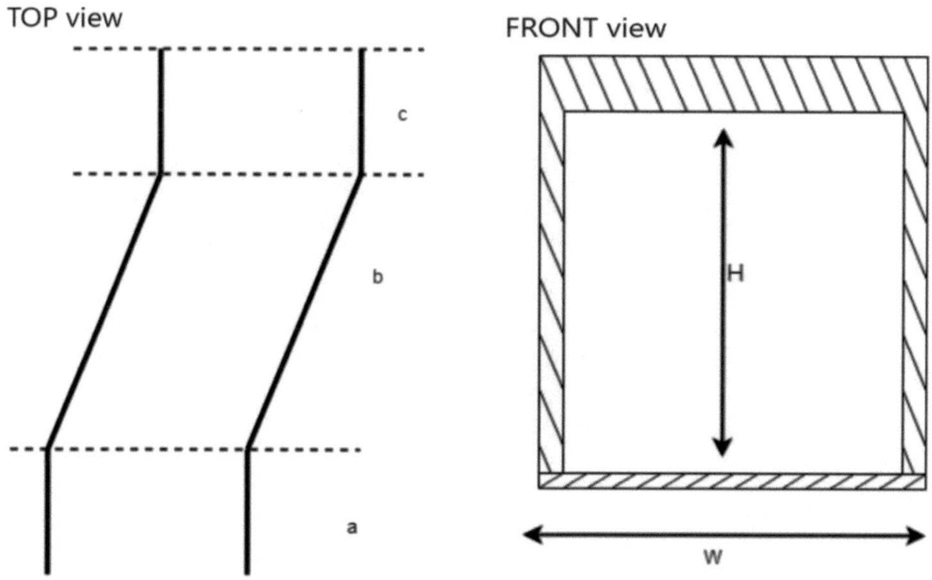

Figure 10.
View of the tunnel.

Test (m)	1°	2°	3°	4°	5°	6°	7°	8°	9°	10°
10 m	**0.701**	0.867	0.383	1.197	0.280	0.981	0.840	1.111	1.269	0.311
20 m	**1.369**	1.150	0.511	0.731	0.403	0.732	1.208	0.820	1.242	0.335
30 m	**1.35**	0.610	0.113	0.689	0.223	0.134	1.383	1.572	1.175	0.301

Table 3.
Absolute error in metres for each different test lengths.

	10 m	20 m	30 m
Minimum	0.280	0.335	0.113
Maximum	1.269	1.369	1.572
Average	0.794	0.850	0.755

Table 4.
Minimum, maximum and average error.

The minimum error obtained for the various ranges of distance tested is consistent with the results obtained in other recent works of the literature [13, 14]. At the same time, if we analyse the average error obtained by performing multiple consecutive tests for each range of distance, it can be seen that an improvement in the visual-inertial system is possible, although the system already guarantees great robustness in operation. An improvement could be obtained by using a different hardware, more performing IMU, and at the same time deepening the aspect related to the calibration of the camera in order to further succeed in decreasing the odometric error during navigation.

After conducting a series of test in facility, we definitively validate the result of the project and the system design in a different real tunnel (**Figure 11**). In this situation it was confirmed that the precision and reliability of the algorithms were enough to allow the system to navigate in total autonomy for at least a stretch of 100 metres.

Figure 11.
Tunnel inspection performed by the UAV during the test of system validation.

4. Discussion and conclusion

The inspection of tunnels and infrastructure for water and hydroelectric resources and, more generally, for any underground work, is essential for the efficient maintenance of the infrastructure itself and provides significant benefits for the rational use of resources. This type of activity collects important information about the current state of consistency of the structure. By highlighting potential or actual failure conditions such as cracks, deformations or other types of problems, it is possible to plan any safety maintenance operations in a timely way. These inspections are now carried out mainly by human operators, with considerable risks to their safety and health at work: claustrophobic, dark and dirty environment.

The idea of this project is to apply innovative techniques, to overcome these problems with the future purpose to ensure greater safety, avoid the inconvenience and risks arising from these activities for the human operator and meet the market needs. As a consequence, a scheduling system has been presented and allows to set different strategies to approach the inspection of the tunnel before starting the mission. Autonomous driving techniques in the six degrees of freedom are developed to ensure the obstacle avoidance in confined space using a simple Lidar sensor. By applying visual-inertial odometry and its fusion with the aid of a Kalman filter, it has been possible the realization of a UAV system able to perform an autonomous inspection of indoor environment like tunnels or conduits.

Although the results shown in this work in terms of robustness and consistency are encouraging, in the future there will be a need to develop advanced techniques considering different scenarios and environments. One possible improvement could be brought developing navigation algorithm based on other types of sensor and using alternative approach. In conclusion, it is central to continue to investigate visual-inertial algorithm since its contribution has proven essential for the robustness, reliability and efficiency of the overall system.

Acknowledgements

The authors want to thank D. Miccone and P. De Petris from WP Web and the Politecnico di Torino Inter-departmental Centre for Service Robotics PIC4SeR, (https://pic4ser.polito.it) for the support and knowledge shared.

Author details

Lorenzo Galtarossa, Luca Francesco Navilli and Marcello Chiaberge*
Department of Electronics and Telecommunications (DET), Politecnico di Torino, Turin, Italy

*Address all correspondence to: marcello.chiaberge@polito.it